Advances in Geographic Information Science

Series Editors:
Shivanand Balram, Canada
Suzana Dragicevic, Canada

Laxmi Ramasubramanian

Geographic Information Science and Public Participation

Springer

Dr. Laxmi Ramasubramanian
City University of New York
Hunter College
Dept. Urban Affairs & Planning
695 Park Ave.
New York NY 10065
USA
laxmi@hunter.cuny.edu

ISBN 978-3-540-75400-8 e-ISBN 978-3-540-75401-5
DOI 10.1007/978-3-540-75401-5
Springer Heidelberg Dordrecht London New York

Library of Congress Control Number: 2009939665

Cover design: Bauer, Thomas

Printed on acid-free paper

Springer is part of Springer Science+Business Media (www.springer.com)

Foreword

Understanding why things are where they are helps all of us understand what can be done to improve the communities in which we live. This is why we use geospatial information technology (to find out where things are) and that is why we have public participation (to understand why they are there – or should not be there). It is through the participative planning process that we can connect the "why" with the reasons to get to "what can be done". In the pages that follow, the reader will learn how geographic information science, public participation, and the planning process can work together to improve our communities.

Dr. Ramasubramanian brings a wealth of knowledge and experience together into a volume that is a significant resource for those interested in participatory planning and, specifically, Public Participation Geographic Information Science – PPGIS. In it, we learn of a vast variety of PPGIS activities and web locations for public access to and use of spatial information and analyses to address public issues. We are introduced to a survey completed in 2007–2008 of the ways in which PPGIS activities are used across the nation and internationally. We also learn, in depth, of three significant case studies involving the public in planning activities using GIS: Boston's South End Planning and Housing Coalition which fought the development of a biomedical facilities building in its neighborhood, the Village of Oak Park (IL) business district development plans, and the Common Ground initiative of the Northeastern Illinois Planning Commission that developed a comprehensive six-county land use and growth management plan. Finally, we learn the value of putting the needs of the public first – not the technology – and to respect the public as responsible stakeholders who are willing to make tradeoffs in balancing their own needs with the need for the community to achieve growth and economic development.

Her framework for creating a "highly customized participatory process" brings it all together, not as a series of steps to be used by planners and elected officials to involve the public in planning issues, but as a set of guidelines that define the conditions necessary for success: the capacity of the participants to become involved in the process; the engagement of those participants in every aspect of the process; the techniques for understanding the data and the views of the participants; and the strategies for deployment of the results of the analyses to the participants.

In the early days of GIS deployment, local governments adopted the technology for their own institutional needs: to become more efficient and effective in fulfilling their mission. We all knew the mission of our institution – and it did not include involving the public in the use of geospatial information or its technology. Over time, however, it became apparent, especially to public planning professionals, that the data and analyses used in local government GIS applications could be used in other ways to help its constituents. It could help developers find new land to develop; to help environmentalists identify geographic areas in need of protection or remediation; and to help neighborhood organizations justify solutions to their problems. How to make that happen, however, eluded public officials who first developed public documents of maps and data tables and later designed web sites that pushed the data and maps out to, normally, an unaware public. (These solutions were designed to reduce the amount of time that a local government employee had to spend with the pubic, thus realizing a governmental efficiency). It took the planning profession – through consensus building, or participatory planning – to actually engage the public in using geospatial data. Dr. Ramasubramanian takes the reader through that development and explains, with examples, how it is possible to successfully implement and modify plans with knowledgeable public input.

A key element in this success is what she terms "Critical Reflective Practice" which strives to seek a balance between activism and reflection (as in research and analysis). The public, she exhorts, must not only take action on proposed changes to their environment, but must also reflect on the issues driving those actions. Such reflection requires information and the ability to analyze it. That is the essence of PPGIS.

There are many groups of readers that can benefit from this text:

Students, researchers, and other scholars who seek to understand the new techniques and critical processes that are necessary to engage the public in issues affecting their lives.

Planning practitioners can benefit especially from her 12 "Guidelines for participatory planning with GIS" because, as she states, "the typical planning practitioner, particularly the middle-level manager charged with ensuring full and active public participation does not challenge the need to engage the public; the question on her mind is not *should I*? Rather it is *how do I*?" Her guidelines are sure to help those practitioners cope with such anxieties.

GIS professionals will get an eye-opener here with the realization that it is not all about GIS and data – it takes an organic process to involve the public in geospatial data and technology so that we all can become empowered to influence decisions affecting our lives.

Finally, the public in general, can get a flavor of how difficult it is for public agencies to successfully elicit the appropriate and rational input necessary to make planning decisions that can benefit individuals in their neighborhoods as well as the city, the developer, and any other stakeholder involved.

Milwaukee, Wisconsin

William E. Huxhold, GISP
Professor of Urban Planning
University of Wisconsin-Milwaukee

Preface

Most first books are autobiographical, and this work is no exception. My interest and commitment to the topics addressed in this book were inspired and shaped by my early experiences as a graduate student at MIT's Department of Urban Studies and Planning. MIT introduced me to the power and potential of new technologies and nurtured my optimism in the positive uses of information technologies. While working with young people in Boston, I became confident that digital technologies like GIS could be used to understand and solve complex social problems. Initially, I believed that the tools and the analyses were central to problem solving. However, during the short time I worked as a neighborhood planner at the Boston Redevelopment Authority (BRA), I learned the hard way that the best tools, techniques, and analyses could not ensure the survival of good planning ideas. One had to understand how planning was actually done, both in the trenches, and in City Hall. From that time, I have been attuned to the complex interplay between the technical/analytical and the institutional/political aspects of day-to-day planning practice. The opening chapter of this book frames some of these complexities.

It's no secret that day-to-day planning can be quite banal – meetings and more meetings, internal discussions, public hearings, glitzy presentations, punctuated with an occasional knock-down-drag-out war of words between irate citizens and hapless professional planners. At the most interesting and productive of these meetings, described in the case studies that form Part II of this book, something very special occurs; ordinary citizens reason together, share data and information, they ask analytical questions of professional planners. Professional planners who participate in these community meetings are respectful, articulate without being condescending, well prepared, and willing to share useful information with the community. It was while attending these meetings that I discovered the real power of digital technologies like GIS. People came together to look at maps and data and the tools, rather than being a hindrance, improved the quality of community conversations. The theoretical discourse about planning practice and the digital revolution, including the promise and some of the intended and unintended consequences of technology adoption for planning are discussed in Chaps. 1 and 2.

In the early 1990s, many community-based organizations (CBOs) were becoming interested in using digital technologies. There seemed an inherent contradiction – watching poorly funded grass roots groups go hi-tech. Intrigued, I conducted a research study through which I learned that community-based organizations (CBOs) were using digital technologies to improve organizational efficiencies, facilitate group processes, and influence negotiations. Since the time I first conducted my research in the mid-late 1990s, there has been a dramatic growth in the number of community organizations that use digital technologies to encourage and support public involvement in planning. The factors that facilitated this growth and the resultant impacts are discussed in Chaps. 3 and 4.

Until recently, mainstream planning practice has had an ambivalent relationship with the notion of public participation. Very often, professional planners are likely to limit discussions about participation to a single empirical question – how many people showed up? "Bums on seats" is a measurable indicator, but very often an irrelevant one. Sometimes, even the people who call the meeting are unclear about the significance of high or low attendance numbers. Did the outreach strategy work well? Is the issue being considered a hot-button issue? Are people participating because they are afraid of change or because they embrace it? The three case studies in Chaps. 5, 6, and 7 provide some answers to these questions.

Fortunately, the culture of planning is changing. One of the first Executive Orders that President Obama signed upon taking office in 2009 concerned itself with transparency and open government. It states:

> "*Government should be participatory*. Public engagement enhances the Government's effectiveness and improves the quality of its decisions. Knowledge is widely dispersed in society, and public officials benefit from having access to that dispersed knowledge. Executive departments and agencies should offer Americans increased opportunities to participate in policymaking and to provide their Government with the benefits of their collective expertise and information. Executive departments and agencies should also solicit public input on how we can increase and improve opportunities for public participation in Government" (Presidential Executive Order on Transparency and Open Government, 1/21/2009).

Even with these hopeful winds of change in the air, one wonders how does Government go about creating an open, transparent, and participatory planning process – is it feasible, is it sustainable, and more importantly, are the decisions made through such open processes fundamentally better than those decisions made by the power elite? Chapters 8, 9 and 10 provide some strategies to implement participatory and open planning as well as strategies for assessment and evaluation.

Who should read this book? Over the past 20 years, I have observed that there is a persistent digital divide in planning research, practice, and education. The worlds of the "techies" and the "activists", i.e., the advocates of participatory planning do not intersect as often as they should. In this book, I strive to bring these two communities and the work that they do closer together to demonstrate the necessary convergences that must occur if we are to improve the quality of civic engagement. I want my research and writing to both usable and useful to practitioners; for this reason, the book blends theory with practice in every chapter. The primary audience for this book is practicing planners, particularly those working within local/regional planning agencies, policymakers, agency heads, public involvement professionals, community organizers and activists. Likewise, researchers working to develop digital tools to support participatory planning and those working to improve the quality of public involvement in planning may benefit from reading about my experiences in the field. Anyone who wants to design and implement participatory planning activities can use this book to consider the pros and cons of integrating GIS and other technologies in these activities.

The book is organized in three parts: In Chap. 1, I establish my case for the need to consider the role of digital technologies in participatory planning and decision-making in the context of planning practice. In Chap. 2, I review the literature about digitally enabled planning practice, with a particular focus on spatial technologies, discussed under the rubric of PPGIS. The myriad contradictions of PPGIS are more rigorously examined in Chap. 3 in order to provide a more coherent understanding and synthesis of the literature. In Chap. 4, the state of PPGIS practice, the results of a national survey of PPGIS activities are discussed to delineate the tensions and disconnects from the world of theory.

Chapters 5, 6, and 7 are case studies, describing events and activities that are quite different from each other. Chapter 6 provides some insights into community-based planning, specifically the work of an organized coalition working in one Boston neighborhood. Chapter 7 describes the experiences of a university acting as a facilitator of collaborative planning in the Village of Oak Park. Chapter 8 provides the tremendous efforts taken by a regional planning agency to engage the citizenry in envisioning the future of the region. In each of these situations, different types of technologies and modalities were deployed in the service of the participatory planning ideal. I selected cases where I had extensive in-depth knowledge, lived experiences, a good understanding of the context of the case itself, and personal familiarity with many of the activities undertaken to achieve project goals, as a researcher, a participant-observer, or as one of the activist-academics engaged in project implementation. One of the shortcomings of this approach is the possibility of bias, of reading into the situation, particular meanings and interpretations that confirm to previously held opinions. To avoid this from happening, I have corroborated my observations as extensively as possible. The conclusions I draw, are mine, supported with evidence drawn from archival materials since many of the participants and initiators of the participatory activities are no longer involved with the projects. In one case, one of the key initiators of the participatory work is no longer alive. I compare and contrast the case studies using a meta-evaluation framework in Chap. 8. My goal is to develop and use this evaluation framework to take a critical look at many more PPGIS implementation efforts, so that we can develop a robust understanding of PPGIS implementation.

The discussion in Chap. 9 discusses the demands that participatory planning approaches place on planners and planning institutions. In Chap. 10, I provide a synthesis and a set of guidelines that can help practitioners create and sustain effective participatory planning projects that incorporate a wide range of digital technologies.

Although I formally began working on this book only 2 years ago, I have actually been writing this book in my head for a long time. A very special thanks to

Bill Huxhold for writing *An Introduction to Urban GIS*; it inspired and opened up a world of possibilities for me. Along the way, I have been influenced, stimulated, advised, and challenged by educators, professional colleagues, community activists and mentors. My thanks to Alex Alexander, Cheryl Ajirotutu, Tom Angotti, Mike Barndt, Mike Batty, MacCanon Brown, Paul Bloyd, Will Craig, Uri Cohen, Judy Colby-George, Andrea D'Amato, Antonia Darder, Ralph Gakenheimer, Joe Ferreira, Roslyn Foskey, Jill Gross, Nacho Gonzalez, Dick Klosterman, Charlie Hoch, Joyce King, Mel King, Leigh Kunde, Melinda Laituri, Ian Masser, Hubert Morgan, Gary Moore, Tim Nyerges, Tom O'Malley, Amos Rapoport, Sue McNeil, Aimée Quinn, Eswaran Selvarajah, A.N. Sengupta, Mike Shiffer, Vonu Thakuriah, Harry Van Oudenallen, Steve Ventura, Doug Walker, Karen Witten, and the late Don Schön, who inspired me and sharpened my thinking about participatory planning practice using digital technologies. The conclusions in Chap. 9 are my own, but they are in all likelihood influenced and shaped by these interactions.

At Hunter College, formal and informal conversations with my students, especially Stephanie Camay, Kate Ervin, Scott Giering, Jason Nu, José Pillich and Deb Stattel have kept me thinking about public participation and the role of digital technologies even when I was not actively working on the book. Richard Amanna, my research assistant helped me immensely in implementing the PPGIS survey discussed in Chap. 4 and in analyzing the data. Brock Doerr, my former student and research assistant provided invaluable assistance by creating publication-ready figures and charts. My academic colleagues at Hunter College, particularly those in the departments of Geography and Urban Planning have encouraged and supported my professional development since I came at Hunter College in Fall 2004; they deserve my thanks and appreciation.

Since I began my academic career, many administrators have helped me with practical advice and assistance; special thanks to Mary Bates, Bob Buckley, Adjie Henderson, Peter Hosking, Albert Schorsch III, and Richard Stayner. The PPGIS conferences I attended and helped to organize under the auspices of the Urban and Regional Information Systems Association (URISA), as well as other synergistic activities sponsored by the National Center for Geographic Information and Analysis (NCGIA) and the University Consortium of Geographic Information Science (UCGIS) have been invaluable in facilitating connections to an active community of scholars and practitioners.

A special shout out to my friends, especially Alberto, Aimée, Brinda, Gowri, John, Len, Mel, Siva, Susan and Sue. Amma, Appa, and Erika, your words of encouragement will always be appreciated and cherished. Last, but not the least, this book is dedicated to my most loyal friend and fiercest critic, Jochen Albrecht. Thanks for reading draft chapters, providing both encouragement and useful feedback, and helping in a thousand different ways so that I could finish this project.

I have participated in many community-based planning initiatives in the United States and abroad. Over the years, I've acquired firsthand experience about the power and value of doing participatory planning using a variety of digital tools. I am delighted to share some of my insights with you. In reading this book, I hope

Contents

Part I
Participatory Planning: Why Does It matter?

Chapter 1
Dilemmas in Contemporary Planning

1.1 Introduction

When I was a graduate student at MIT, I signed on for a part-time job. A MIT researcher.[1] I knew was in the process of perfecting a new piece of software called City View/Town View. The program was designed to assist novice users in describing their own neighborhoods using an electronic story board. The program was not a complex one; it was essentially a customization/adaptation of Apple's HyperCard.[2]

Part of my responsibility was to assist the researcher in beta-testing the software by engaging middle school kids from Roxbury[3] in the use of the tools. In practical terms, this meant that I had to work with the students as they attempted to describe their neighborhood using pictures (images) created using the software. The seven young people I worked with that summer were bright and motivated, in part, because they were being paid a small stipend[4] to participate in the program and because their work had a "cool" factor – they got to hang out in their school's computer lab and learn new software in a relatively unstructured way.

The researcher, understandably, was more focused on elements associated with the software application, considering issues such as usability, the functionalities of various menus/buttons, and on generating a work product that could show-case the capabilities of the tool, rather than becoming overly involved with student learning goals. The task of motivating the students to do the actual work of creating a physical representation of their neighborhood fell to me. This was no easy task for a newly-minted city planner without any formal training in working with teenagers.

"Think of landmarks[5] in your neighborhood", I remember saying, and was met with blank looks. "Churches", "important buildings", I persisted. "There's nothing like that around here", came the nonchalant response, contradicting objective reality. Not making much progress, I tried another approach; "How would you give me directions, if I were walking to your house from school?" Immediately, they brightened up and started drawing.

Over a period of time, using the drawings as a guide, we began to have very interesting conversations – using the map of the park; one boy told me his story – "This is the park that I have to cross every day". When I asked him to trace the actual route, I noticed that the line did not cut across the park, but went around it. Upon further inquiry, he noted, "I always walk on this side of the park (pointing to it on

L. Ramasubramanian, *Geographic Information Science and Public Participation*, Advances in Geographic Information Science, DOI 10.1007/978-3-540-75401-5_1,

the drawing) because it's safer. There are older kids who hang out on the other side, and sometimes there are drug dealers." And so it went. Over the next few weeks, I learned a great deal about how these seven young people experienced life in their neighborhood and community; what they liked about their built environment and what they didn't, what they wanted to preserve, what they wanted to change.

My work with the youth was by no means an ethnographic study where I spent months and years with the same group of people, living and working among them. Yet, I feel privileged and lucky to have had these intimate conversations at the beginning of my professional career because it made me realize very clearly that conversations are at the heart of building community, and absolutely essential if we are to see the world through another's eyes. In my work with the youth that summer, if I had focused exclusively on the task at hand – the creation of visual representations, I would have missed so much.

In one of these conversations, I realized that many of the kids had never left their neighborhood. So, instead of diligently working on the tools, we jumped on the subway and did a bit of exploring. I remember taking the students to the offices of the Boston Redevelopment Authority for a tour, and to look at the scale model of the city that was on display. They loved the model, orienting themselves and their neighborhood in the context of the large swath of the city. "This is where we are"; "Here's Washington Park". Soon we were talking about our futures – college, work, life. We were planning.

Back in the classroom, our conversations gradually moved to the next level – from describing problems to a careful consideration of how these problems could be solved. One young girl spoke passionately about the vacant lot next to her home. "It's full of weeds and there's trash all over the place". When asked, what she would do to fix things, thinking about her little sister, she said that she would like to turn the vacant lot into a nice park for little kids, who didn't have a place to play close to home under adult supervision. With some encouragement, she began the task of designing her park – eventually getting down the nitty-gritty details like designating furniture and play equipment for the space. By the time, she was finished, she was convinced of the viability of her plan and was ready to talk to as many adults as she could, so that she could get her plan implemented (Fig. 1.1).

Another student observed that youth in the school were often unaware of activities and opportunities that were available to them during after-school hours or on weekends. He suggested that many kids "hung out" on street corners because they didn't have many alternatives. After some brainstorming with his peers, he decided to create an interactive map that showed locations of various youth-serving agencies. When you "clicked" on a particular location, the user would be able to view information about opening hours, and types of activities. His end goal was for the map and the underlying database to be available to the students on the school's computers.

The project wrapped up with a final presentation where adults, fellow classmates, school teachers, MIT professors, and invited guests from the City celebrated the students' accomplishments that had resulted from a 6-week after-school program (Fig. 1.2).

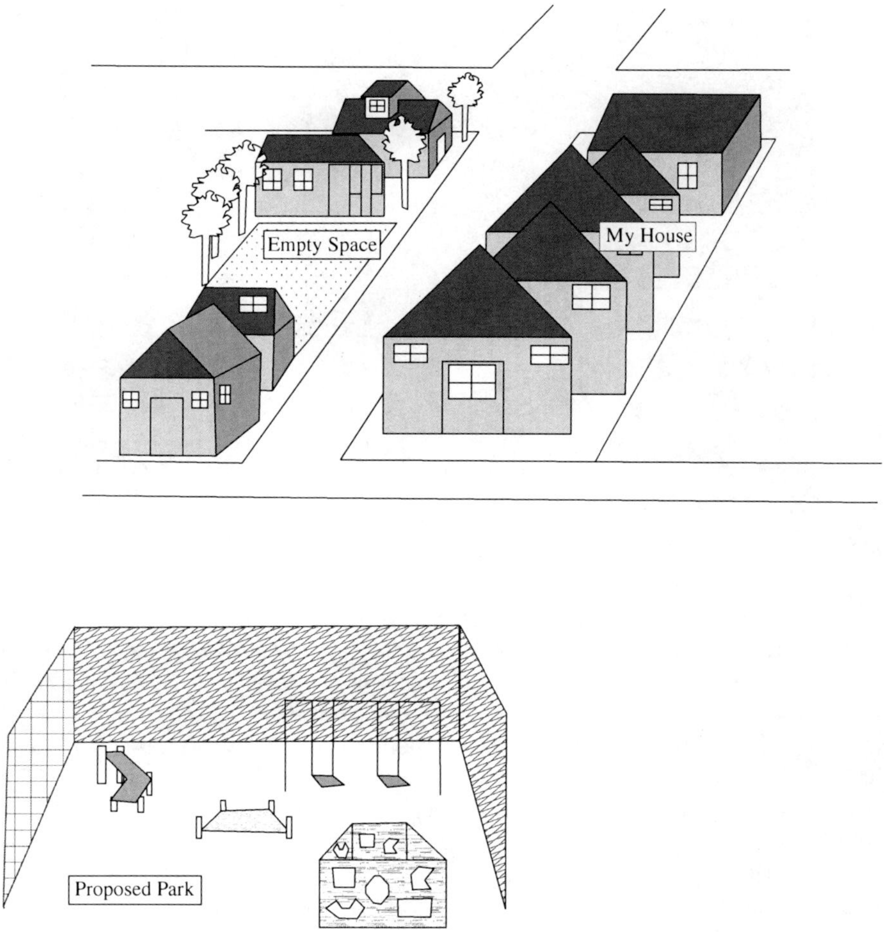

Fig. 1.1 Georgia's Park on vacant lot

I was aware that the young people had changed over the course of the short time I'd spent with them. In particular, I was struck that they seemed to express a more positive attitude about their neighborhood and community (their lived environment) and their own role in shaping its future. As we went our separate ways, I got to thinking about "what worked and why?"

- What did this experience teach me about the nature of public participation?
- What did this experience teach me about the use and the limits of using technologies to facilitate participation?
- What had I just learned about the nature of planning itself?

These three questions and the complex answers to them shape this book.

Fig. 1.2 Author with youth

Let's discuss these questions, one at a time, beginning with what appears to be a benign and easily understood word – participation. At a very basic level, participation means to be a part of a larger whole, to share and engage, offering contributions and reaping benefits. In a very basic way, participation is always a two-way street and an interactive experience. It is useful to keep this definition in mind, when we investigate how public agencies conceptualize participation in our case study narratives.

Participation, specifically, in planning and decision making, the subject of this book, is something that is more easily accepted in word than in deed. There is a simple reason for this – elites, and most professional planners and policymakers of the world are elites, the world over believe quite fervently that they know what is best for the people and the community they are attempting to plan for, whether it's the local neighborhood association or the nation state itself. These beliefs are also shared by many in marginalized communities; I have been chided more than once for asking for opinions and perspectives of local residents by community leaders, "why are you asking us; haven't we hired you because you are the expert?"

In attempting to unpack the complexities of participation in planning, consider that the young people that I introduced you in previous pages were most motivated and engaged when they were "teaching" me (a grown up) about their neighborhood, and the effort they invested in carefully detailing their experiences with complexity and nuance was a first and necessary step in their (and my) learning process. The information they shared with me about their neighborhood and community was unique and special, and was definitely not something that would be easily accessible to an outsider. As we wandered the city together, some barriers of race and class seemed to fall away, even if they were never completely erased.

My work that summer was very different from the planning I'd studied in grad school – it was fun! I was the student and the young people were the teachers. By surrendering some of my own authority, the kids and I had both changed how we viewed our world. But, what does this have to do with "real planning"?

At its simplest, planning is a way to envision the future. In an ideal world, planning seeks to create "better" futures for all citizens. Planners have long allied themselves with social reformers; many late nineteenth and early twentieth century planning initiatives were situated in populous and polluted cities including London, New York, and Chicago (Hall, 1996). Many of these reformers placed great emphasis on meeting the needs of economically and socially vulnerable populations.

Contemporary planning is about developing affordable housing, creating well-paying jobs, ensuring safe neighborhoods and healthy communities. While some of these goals are achieved by creating a range of mechanisms (e.g., legislation, guidelines, and new institutions with authority to review and evaluate both processes and outcomes) planning in the United States is also often closely associated with dramatic changes to the physical landscape. Recent examples of such physical planning initiatives in the United States include the Central Artery/Tunnel project in Boston,[6] or the Hope VI initiatives sponsored by the US Department of Housing and Urban Development.[7]

The United States is unique among other industrialized and modern nations in that it has no federal department of planning. Planning activities are split across a plethora of agencies and branches of government. At the local government level, planning includes: comprehensive planning, planning for affordable housing, economic development planning, urban design, zoning, and growth management. At the same time, the federal government has created a variety of agencies (e.g., Environmental Protection Agency), and has passed legislation (e.g., legislation to establish Metropolitan Planning Organizations or MPOs) to encourage municipalities to work collaboratively across regions. It is useful to remember that many planning projects in the United States are undertaken by private individuals or private business consortia. Philanthropic and religious organizations, non-profit groups, and individuals are also actively engaged in the business of planning.

Planning practice in the United States is shaped by particular conventions, i.e., institutional and political norms, protocols, methods, and systems. Many of these protocols use analytical methods to make intelligent judgments about the future. For instance, planners frequently use socio-demographic analyses to answer questions such as – Who lives here? What kind of work do they do? Do they need new schools? Will they need new schools in 10 or 20 years time? For each question, there is data that is assembled and analyzed – thus a planner may speak about the need for new housing depending on the population projections for a particular area, taking into account fertility trends, migration to and away from the area, the number of available housing units and their costs, and the kinds of jobs that are available/likely to be available in the future.

So, let's revisit my experiences planning with the young people in Boston. If I'd relied on official statistics to help plan for the future of the neighborhood, I would have gathered lots of data about material and social deprivation (number of children who were qualified to receive a free or subsidized school lunch, the number of homicides in the past year, the number of individuals with a college degree, the number of unemployment claims in the past 6 months and so on). These statistics would have told me one kind of story, the story you'd expect to find in a poor neighborhood.

There is another kind of story; the story that the kids told – of caring, simple hopes, fears, and struggles. The intricate patterns of living and experiencing the neighborhood – both its dysfunction (no child should have to re-route their walk home to avoid being intimidated by a gang) and its social cohesion (the care and concern that the teenager Georgia expressed for her younger sibling and for other little kids) would have been rendered invisible in a flurry of numbers and statistics. Collectively taken, the statistics would have made it impossible to envision a hopeful future for the youth in the neighborhood and all the more easier to write off as a lost cause.

I began to wonder whether it was possible for experts to plan for the needs of a particular community without actually speaking to the folks who lived there. More questions began to emerge: does all planning have to involve this kind of in-depth consultation? Who should be consulted? Is consulting kids even relevant? Are these consultations merely feel-good exercises designed to assuage middle-class angst? As a scientist and thrifty tax payer, I asked myself, whether the final recommendations/plans were any better because of all the community consultation. I address these and other questions in later chapters.

I have stated that planning begins with the articulation of a vision. In order that the vision is anchored in some reality, it must include some facts, figures, and data that have to be organized and codified to make a convincing argument. When Georgia proposed that a park be created to replace the vacant lot in her neighborhood, she was considering the needs of a particular demographic; young children and their parents who preferred a play area that was safe and easily accessible. I hypothesized that the digital tools she used helped her to organize her narrative to create a strong case, a proposition I chose to explore further, over the years.

Notice, I haven't said much at all until this point about the technology, the intermediary that we used to facilitate our conversations. The technologies are significant and deserve to be celebrated in their own right, but for now, let us consider specifically those features/capabilities of the tools that proved to be most valuable in facilitating dialogue and collaboration.

First and foremost, the software program/application was simple to learn and use – probably one of the fundamental requirements of any software program that is intended for use in a participatory setting. Second, the software allowed disparate pieces of data to be stored and organized in separate "cards" that collectively became a searchable database "stack". The data on each card could be organized and re-organized to create different types of presentations. For instance, a series of user-generated images could be used in turn to describe different users' routes from their home to school, the location of landmarks, or areas of the neighborhood in need of revitalization. Finally, the program allowed users to create hyperlinks, allowing the user to retrieve and display additional information at the click of a mouse.

All this may seem very tame to the reader at the present time, but back in 1991, these technologies were quite sophisticated. One of the benefits of using the technologies to engage youth was that they were more intrigued and interested in getting involved in our project because the technologies represented something new and different.

I was left with more questions – could these great conversations have occurred without the use of these technologies? As a trained architect, it seemed natural to use conceptual drawings, maps, and diagrams to enable communication but the precise contributions of these digital tools were still unclear to me.

1.2 Participation and Planning: An Uneasy Relationship

Participation in planning is something that is more easily accepted in word than in deed. Therefore, it should come as no surprise to learn that public participation in planning decisions was limited to the power elite during the first half of the twentieth century, although the generally reform-minded planners of this era believed that they were acting in the public's interest (Hall, 1996). Despite these good intentions, the planning profession has had a long trajectory of developing and supporting comprehensive planning initiatives that have gone terribly awry. The destruction of neighborhoods and communities unleashed by the highway building programs and the urban removal programs of the 1950s and 1960s created a justifiable mistrust about professional planning initiatives. The backlash against comprehensive top-down planning of the 1960s and 1970s helped spur the development and acceptance of the culture of citizen participation in planning (Davidoff, 1965). The apparent arrogance of professional planners, those who sought to define vibrant neighborhoods and communities in bricks and mortar terms alone, angered citizens already energized by the *zeitgeist* of the civil rights struggle. The 1960s were a time when ordinary citizens organized and mobilized to challenge the professional wisdom of significant planning decisions (King, 1981; Medoff & Sklar, 1994).

Since the 1960s, when the federal government included "citizen participation" as a requirement in its antipoverty programs, citizen involvement in professional planning efforts has been *de rigueur* (Hoch, 2000). Furthermore, direct participation in governmental decision making is viewed as the cornerstone of a vibrant democracy (e.g., Barber, 2004). Yet, it is a concept that seems to been accepted more in theory than practice. Planning practice interweaves conceptual ideals of public participation within existing decision making structures, thus resulting in some enduring dilemmas for practicing planners.

1.3 The Core Dilemmas

As a practice oriented discipline, planning is incredibly self-conscious and analytical about its role and purpose. A large body of theory, often called "theory in practice" has been assembled to discuss the core dilemmas that affect all planning endeavors (Schön, 1983). Each dilemma discussed below is linked to some aspect of public participation:

1.3.1 Framing Planning Problems

Framing a problem has a powerful impact on the solutions that are proposed. Schön and Rein (1994, p. xii) propose that institutional action frames are "beliefs, values, and perspectives held by particular institutions and interest groups from which particular policy positions are derived." While rational planning is successful, in part, because it helps integrate data and analysis to establish causal chains, it also is spectacularly unsuccessful when it is required to integrate non-quantifiable, non-economic models of cause and effect, often hidden within institutional action frames.

Community activists, in particular, have long known that it is near impossible to shape outcomes of some planning studies, because they are framed in ways that can result in outcomes that are only suitable to the framers. For example, in 1960s, when urban renewal was at its peak, the discussions about the need for urban renewal were cast (framed) as problems of poor housing and living conditions (sub-standard and dangerous structures, health and safety of residents was at risk because of living in over-crowded conditions), wherein the only plausible solution was to remove the decrepit housing stock and replace it with new, presumably, better quality housing. However, intangible qualities such as sense of community could not be factored into any analyses, given that the problem focused exclusively on the built environment. This point was brought home to me during an interview with a long-term resident in the South End of Boston who said:

> ... the people from Harvard and MIT didn't know any of the people that lived here, but they looked at all the buildings. Oh! They're all rotting away, so let's just knock 'em down and build something else [better] (Resident interview, 1996).

1.3.2 Determining the Locus of Planning Authority

It is often argued that the rational planning model survives because it "appears to provide a strong rationale for professional expertise" (Hoch 2000, p. 23). However, since the 1960s, when the first challenges against expert-driven, institutionally mandated planning occurred, community activists and citizens have consistently challenged the authority of professional planners.

Advocacy planning emerged because savvy community activists and organizers recognized that professional expertise was often used to thwart the challenges posed by average citizens. As a response, advocacy planning as practiced in the

1960s championed a legalistic approach (akin to providing poor/indigent citizens with the services of a public defender). In this model, "progressive" expert-planners argued against other planners working for city government on behalf of beleaguered "naïve" members of the public (Davidoff, 1965). Advocacy planners used the language of expertise to challenge unspoken assumptions, many times revealing inaccurate and sloppy analyses. These activist-planners drew attention to the social issues that were being ignored because of the emphasis on the built environment.

Participatory planning, as it evolved in the 1980s and 1990s validates the voices of experience, that is to say, the voices of those who are directly affected by particular planning decisions (Freire, 1970; Gaventa, 1993). Both participatory and advocacy planning have made some significant inroads in shaping conventional planning processes. Presently, all planning activities that have the power and legitimacy of the State associated with them include some opportunities for citizen review. However, the essential dilemma – the legitimacy of professional planning continues to be contested terrain.

1.3.3 Defining the "Public Interest"

The United States, because of its history, and as a relatively young nation, has always been reluctant to subsume individual rights and primacy of private property ownership under law or legislation. Land use (a designation determining the type of use such as residential or commercial) and zoning (a designation determining the height and mass of a building) requirements are often used to implement planning decisions. For the early social planners, zoning was necessary to protect the general public (ensuring light and air, safe working conditions, reduction of overcrowded housing conditions, etc.) against unscrupulous profiteers. For example, the roots of zoning law established in New York[8] were designed to prevent individual property developers from designing buildings that would block access to natural light and air, thereby affecting the quality of life of residents in adjoining properties.

Later on, these laws were expanded when the Village of Euclid, Ohio, zoned land to preserve community character by imposing use and density restrictions. The ensuing 1926 Supreme Court Case (Euclid v. Ambler),[9] upheld the rights of the Village of Euclid. The ruling established the need to protect the public interest against individual owners or developers who, in their desire to maximize profits, were likely to ignore concerns about health, safety, or quality of life concerns. In the last half-century, zoning has become a powerful instrument to shape and guide development and determine the visual character of a neighborhood or community. Regretfully, zoning is a blunt instrument and a less than thoughtful application of zoning laws has many negative consequences.[10]

Eminent domain, the taking of private land for public purposes, by government is highly controversial in a society that places great value on individual property rights. It has often been used for the development of large scale infrastructure or transportation projects that require many acres of contiguous land for development. More recently, in the 2005 Kelo case (Kelo v. City of New London),[11] the United

States Supreme Court ruled that the community's desire to support economic development justified the taking of private land using the principle of eminent domain. As planners strive to represent the needs of the many, include those who are not present (under-represented populations and future generations), the concept of the "public interest" continues to be negotiated and re-defined to suit particular situations and contexts.

1.3.4 Management of Participation Within Formal Processes

While citizen activists and special interest groups vociferously clamor for increased opportunities for participation, there is a growing and uncomfortable realization that citizen participation has become a series of formalized bureaucratic rituals (e.g., designated periods for public comment) that are ineffective and sometimes counter-productive (Innes & Booher, 2004). Professional planners observe that public participation as currently managed sometimes undermines their professional expertise, reducing them to "glorified event planners".[12] Planners working in public agencies continue to be uneasy about opening up professional planning processes to the general public. Carp (2004, p. 242) explains these attitudes thus; "public participation costs time and attention; and to the extent that it introduces political and interpersonal complexities into decisions, it compromises planners' autonomy and efficiency". In addition, planners are also concerned about raising expectations among citizenry by promising more control over a project that can realistically be delivered.[13] Finally, planners working for government agencies are also ambivalent about citizen participation because their counterparts in the community (advocacy planners working with/for communities) tend to perpetuate an adversarial relationship with them.[14]

1.4 Planning and Technologies

Planning practice, particularly in the early 1960s was enamored by computers – computers were associated with order and reason. At this time, planning was considered an "applied science" and information technologies were used to support a rational and presumably value-neutral planning process. Although the gleam associated with computers had largely worn off by the 1990s, the declining costs of personal computers; the use of graphical user interfaces that allowed users without sophisticated programming skills to operate software programs; the increased availability of "official" data such as the US Census; and the emergence of the World Wide Web collectively re-energized debates about the potential benefits and constraints of using digital technologies to support planning activities.

Digital technologies offer the potential for great public benefit in the areas of education, health care, business, commerce, environmental management, and community revitalization. Recognizing their potential, the United States government, corporations, and other not-for-profit entities all invested heavily in these technologies throughout the 1990s. In their comprehensive study of relationships between information technologies and city development, Stephen Graham and Simon Marvin (1996, p. 336) suggested that digital technologies "facilitate control over space for powerful groups, while creating new physical and electronic ghettos for marginal, low income and disenfranchised households". While the observations of Graham and Marvin are not always directly supported by empirical evidence, we can speculate on the validity of their observations by drawing an analogy from significant networks of an earlier generation – highways and freeways. The absence of efficient and accessible transportation links prevented some citizens, those without private transportation for example, from gaining access to employment and economic development opportunities outside their immediate neighborhoods, thus contributing to the creation of economic and social segregation. The extension of the highway analogy to the "infobahn"[15] is, by no means, a stretch of the imagination.

From the beginning, it seemed obvious that digital technologies and the ensuing decisions made by using them were going to directly or indirectly affect the lives of ordinary people. This is an age when people tele-commute to work to avoid sitting in traffic, and global business outsourcing makes it possible for restaurant reservations in New York to be managed from Bangalore, India. Our hyper-connected citizens use their mobile phones to write emails, send text messages, "twitter", update their web profiles, and use the built-in global positioning systems (GPS) to find their way around town. The dreams of the true believers like Negroponte and Bill Gates have come true! And yet, there is a persistent problem that refuses to go away – digital technologies have not completely erased the old barriers of race, class, gender, and age. Contrary to expectations, digital technologies have further exacerbated the problems faced by marginalized communities. Nowhere are these tensions more visible than in the city development – the domain of planners.

Digital technologies have empowered some – the previously privileged are even more so, while those without education, technical skills, and access to the possibilities that technologies can offer are left out in the cold. Consider this scenario; you are in a new city and have lost your cell phone. Good luck trying to find a working public phone; it would appear that city planners and phone companies throughout these United States have decided that there is no need to provide for public telephones anymore. Perhaps they expect everyone to carry a cell phone or a laptop that can be connected to a wireless network. This experience may be a minor inconvenience for some, but a real hardship for others. Recently, Google launched a service called Grand Central, now Google Voice, that tries to address some of these difficulties. The new service gives users "one telephone number for life", and enhances the functionalities of mobile phones including the ability to retrieve voicemail messages by phone, email, or online via the web.

An offshoot of Grand Central is Project CARE (Communications and Respect for Everybody). This service offers, free of charge, a local phone number and voicemail box for life, for members of the homeless community. This project is being pilot-tested in San Francisco before it expands nation-wide. According to Project

CARE, having a consistent and reliable phone number will "provide homeless clients with a link to the real world and an ability to connect". Despite its laudable goals, Project CARE still reveals the fundamental disconnect between the capabilities of services offered by digital technologies and the lived experience of homeless individuals in the "real world".

High rent districts all over the United States have the advantages of 24/7 physical and virtual connectivity. At the same time, every major US city has many underserved neighborhoods, where living and working conditions are less than favorable. Here, people with limited formal education, live in sub-standard housing and work in low-wage jobs far from home. Consider a scenario where a young mother of two is looking for a minimum-wage service sector job. Most jobs that she seeks are probably far from where she lives – her commute will be long; she most definitely cannot tele-commute. She needs to find affordable transportation options and dependable daycare – there is very little that she can gain from getting onto a computer, even if she found one to use. At the most, she could find out bus routes and schedules, but the service on the ground is unlikely to be prompt or reliable.

To summarize, when digital technologies are discussed in the context of planning, the first issue that must be considered is the persistent digital divide between information haves and have-nots. But having access to information is not a panacea by itself and should not be seen as a substitute for participation in planning activities. The case discussed in the next section will clarify this distinction.

1.5 The Way Forward

Let's take a look at a landmark case involving the dramatic use of digital technologies in planning. In April 1995, the American Family Mutual Insurance Company in Milwaukee, Wisconsin settled a discrimination case by agreeing to invest 14 and a half million dollars in the central city. The plaintiffs in the case, the National Association for the Advancement of Colored People (NAACP), argued that the company was under-serving the predominantly African American community residing on Milwaukee's north side. While the case never went to trial, the plaintiffs and their attorneys had gathered a significant volume of statistics and analyses to support their claim. Through this settlement, the community received substantial financial compensation for programs developed to subsidize interest rates for home purchases and home improvements. The settlement provided financing for emergency home repair assistance, home ownership counseling, in addition to providing financial relief for individual victims of discrimination.[16]

Both parties used geographic maps of the area, one of which reflected the number of actual policies issued (Fig. 1.3).

Although the insurance company argued that it had sold many policies to homeowners in majority African American zip code zones, the plaintiff's analyses demonstrated that the company's best policies were clustered in largely white census tracts. Mapping data by address and viewing the display at the scale of census tracts

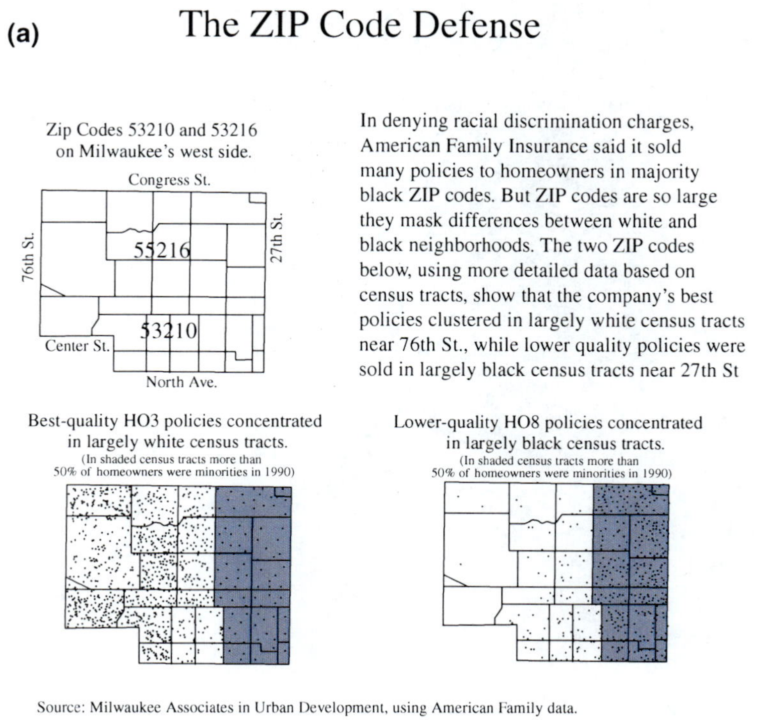

Fig. 1.3 (a) High-tech redlining

provided an understanding of the problem not previously revealed through discussions and analyses based on zip codes, since zip code zones tend to be large enough to mask differences between predominantly White and Black neighborhoods.

In this case, we have to ask: Why did the community groups bringing the complaint against the insurance company use digital technologies? What prompted them to use maps in particular? How and why did they decide to look at the same information at two different units of analysis (census tracts versus zip codes)? Did they ask the questions and analyze the problem themselves, or did they leave the analyses to consultants or technicians outside the organization?

The NAACP v. American Family case is just one of many examples that demonstrate how access to relevant information played a vital role in identifying issues and placing them within a problem solving framework. In addition, the case demonstrates that digital technologies played a significant role in supplying the comparisons and analysis of trends that made it possible to establish discriminatory behavior against the insurance company.

I propose that the success of this particular lawsuit was largely due to the ability of community activists who were able to link the power of the digital tools with more traditional community mobilization strategies. The tools helped to give

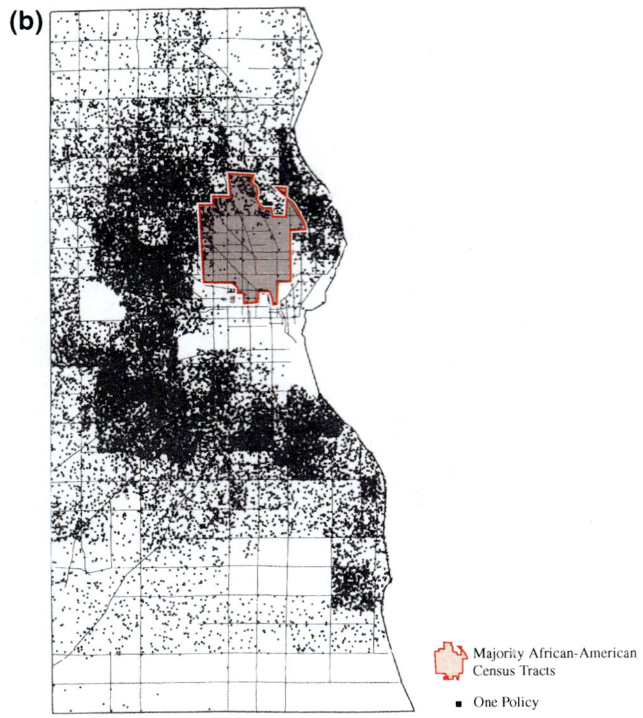

Fig. 1.3 (b) Redlining in Milwaukee

voice to latent concerns that the residents had – where quantitative/visual evidence complemented and affirmed experiential knowledge. Today, the work of these community activists, specifically the use of digital technologies within a community organizing framework would be labeled Public Participation GIS or PPGIS. The early origins of PPGIS were focused on harnessing the capacities of GIS to serve community interests; it was simply another tool in the advocacy planner's tool box.

Planning in communities that are ravaged by material deprivation takes more than data – it takes hope and courage. Regardless of whether we are technophiles or technophobes, theoreticians or practitioners, community activists or academics, we should be asking serious questions – Will the use of digital technologies in planning make decision-making processes more transparent and accountable? Will the use of these tools subordinate common sense, further alienating the general public? Will digital technologies support or hinder participatory planning activities? Are digitally enabled participatory planning activities sustainable in the long run? In Chap. 2, I review the academic literature in order to answer some of these questions.

Notes

1. Barbara, L. Barros was the President of a Boston-based company called *StrataVarious* Inc. that developed HyperMap atlases, where users could create customized maps.
2. HyperCard circa 1990, last sold in 2004 was an Apple product that allowed users to create customized databases.
3. A more detailed description of the project and program can be found in King and Ramasubramanian (1997).
4. Under the guidance of Mel King (Professor Emeritus at MIT and presently Director of the Technology Center @ Tent City) and Prof. Antonia Darder (now Professor at University of Illinois Urbana-Champaign), I wrote my first research grant proposal to the HayMarket Peoples Fund to secure the money for the students' stipend.
5. I was using conventional professional/expert language of urban design that included the well-established nomenclature to describe a city, paths, edges, nodes, landmarks, and districts. For more information, read, Lynch (1960).
6. The Central Artery/Tunnel project (aka the Big Dig) rerouted and submerged a 3.5 mile section of an interstate highway running through the heart of the City of Boston.
7. Hope VI is an initiative of the US Department of Housing and Urban Development that sought to revitalize the worst public housing projects, by replacing high-rise buildings, with low-rise mixed income housing developments.
8. Joseph (1930).
9. US Supreme Court (1926).
10. See Jackson's influential book on suburbanization Jackson (1985).
11. US Supreme Court (2005).
12. Interview with practicing planner by author, November 2007.
13. Civic Alliance (2002).
14. Angotti (2007).
15. The word *infobahn*, borrowed from the German *autobahn* is more eloquent than the term *information superhighway*. Both terms evoke images of high speed and efficiency.
16. The NAACP v. the American Family Mutual Insurance Company Summary of Proposed Settlement Agreement. Copy of agreement furnished to the author by the American Civil Liberties Union of Wisconsin. In addition, the case was widely discussed in local newspapers (Ramasubramanian, 1995).

Values
conservation

Framed
Socially constructed

Chapter 2
The Digital Revolution

2.1 Introduction

In the last chapter, I described a 1995 settlement negotiated between a powerful corporation (an insurance company) and relatively powerless homeowners. In this classic David vs. Goliath scenario, against all odds, the homeowners emerged victorious. Digital technologies, data, and information, anchored by sustained community activism influenced settlement negotiations. In this chapter, I propose that the roots of this particular type of digitally mediated community advocacy were established a long time ago – in the late 1960s in fact. In this chapter, I trace the history of the digital revolution that occurred in parallel with the more prominent civil rights revolution in the United States.

I propose that the unique circumstances that have shaped digitally mediated community advocacy in the United States must be understood in order to better understand its multiple dimensionalities. But before we get to that, I want to remind you about few basic assumptions that I've made already. These assumptions may be pretty obvious; nevertheless it is good scientific practice to make one's own assumptions and beliefs as explicit as possible.

The entire discussion in Chap. 1 is centered around three inter-twined themes, planning, participation, and digital technologies. This leads us to our first assumption – that there is something valuable to be learned in investigating the nature of public participation and the role that geo-spatial technologies can play in facilitating or hindering such participatory activities. Implicit in this discussion is the notion that the use of digital technologies are "socially constructed". This means that all technologies, in our case, information and communication technologies are not imbued with inherent power. Rather, the manner in which the technologies are designed, deployed, and disseminated has significant social, cultural, and political consequences that can be positive or negative. For instance, a cell phone can be used as an instrument for simple communication, for surveillance (when you enable the phone's GPS capability to track down the location of your child), or as a reward (when a school allows its high achievers students to use their cell phones while in school). As the I-Phone ads promise, "there's an app for just about anything"! The ad draws our attention to the different ways in which software developers and

L. Ramasubramanian, *Geographic Information Science and Public Participation*, Advances in Geographic Information Science, DOI 10.1007/978-3-540-75401-5_2,

users have created and adapted a basic communications device to accommodate both practical and emotional needs.

Another assumption is that digital technologies are transforming the nature of contemporary planning practice, and these transformations can be understood only in observing planning where and when it happens. Planning practice is governed by national, regional, and local cultural and political contexts of decision making. These contexts shape how technologies are deployed in the service of fostering public participation. A corollary to this observation is that the adoption and use of technologies is a dynamic process – i.e., planning practice is affects the design and development of tools and new tools help re-energize planning practice. For instance, GIS has now become part of the strategic planning playbook of many community organizations (CBOs) partly as a reaction and adaptation to the use of GIS by planning agencies. At the same time, CBOs have also adopted GIS because it provides an edge (or is perceived as providing an edge) while negotiating with more powerful entities.

Finally, for the purposes of this book, I am proposing that geo-spatial technologies are a subset of a larger set of digital tools and dependent on a global data infrastructure. The development of GIS tools and processes cannot be understood without understanding the digital revolution. Keep these assumptions in mind; we'll return to them later, as we attempt to draw some general conclusions about the three core topics that will focus our discussion – technologies themselves, public participation, and the day-to-day practice of planning.

2.2 The Digital Revolution

There are countless narratives about information and communication technologies, and each narrative has its own starting point. For example, one could begin by focusing on the Pony Express, a cutting edge form of information and communications technology in its day. Suffice it to say that technological innovations often occur in relatively short spurts of intense development, and one innovation often triggers another. For instance, in an earlier era, at the turn of the nineteenth century, the processes of city development were accelerated and completely transformed by the introduction of the new technologies of the day such as the telephone and the telegraph.[1] This is a trend that continues even now, as the use of digital technologies permeates and shapes different aspects of city development (Audriac, 2005). Consider that when you book your hotel room in any modern city this year, you are likely to take availability of free high speed wireless Internet service for granted. This service would have been deemed a luxury, or simply not available even a few years ago.

I've made a choice to begin my discussions from a rather turbulent year in American history – 1968. It serves to establish a linkage between technological developments and societal developments in the United States over the past 40 years. In examining changes in information and communication technologies over this 40 year period, let's first look at the early years, 1968–1978.

2.2.1 Developing Technologies (1968–1978)

It is often stated that the key ingredients of the digital revolution are the personal computer and the presence of the Internet. For most of us, both these developments are a 1990s phenomenon. However, the seeds for these innovations were sown in the 1960s when the forerunner to the modern Internet was created by the Department of Defense through its ARPANET program. The first personal computers were introduced in the early 1970s. Subsequently, Graphical User Interfaces (GUIs) were developed. Early computing systems were clunky, unwieldy and expensive. They were also developed to serve the needs of powerful investors – almost every IT innovation of the 1960s was driven largely by military/national defense goals. In large part, the focus at the time was about efficiency and accuracy, that is to say, "how can we get computers to do things faster and more accurately than humans can"? In this "trickle down" model that would subsequently be mimicked by every new innovation, the development of information technologies moved from defense and national security arenas into the realm of public administration. Among the early adopters were governmental agencies that began to "computerize" their records in order to automate and expedite internal institutional obligations such as payroll and inventory management as well as customer-oriented services such as billing.

Around this time, the first Geographic Information Systems (GIS) emerged – GIS was a term developed by Roger Tomlinson, who designed the Canadian Geographic Information System (Tomlinson, 1987). Other individuals and agencies developed tools, and protocols to facilitate rudimentary spatial functions.[2]

In the 1970s, IT development and GIS development occurred separately in local governments; GIS was sometimes treated as a stepchild of larger IT initiatives. An exception to this norm was the development of the United States Census, which embraced GIS and developed the early protocols and standards to create street network and address files that allows computers to automatically search for and locate a particular address on a digital map.[3] Another early adopter of GIS was the US Geological Survey, the nation's largest civilian mapping agency established in 1879. Working in partnership with the private sector, the USGS has consistently created and assembled mappable data about a variety of urban and natural attributes (Fig. 2.1).

2.2.2 Developing Software, Data, and Applications (1978–1988)

In the 1980s, personal computers and software to serve applications needs began to emerge. Although these computers and software programs still served an elite market, users were able to acquire and use these technologies to serve everyday needs. Spreadsheets, relational database systems, and word-processing applications became popular around this time. At the same time, GIS users too continued to evolve and grow – adopting and adapting other emergent technologies such as the use of remotely sensed data. The major drivers of GIS applications development in

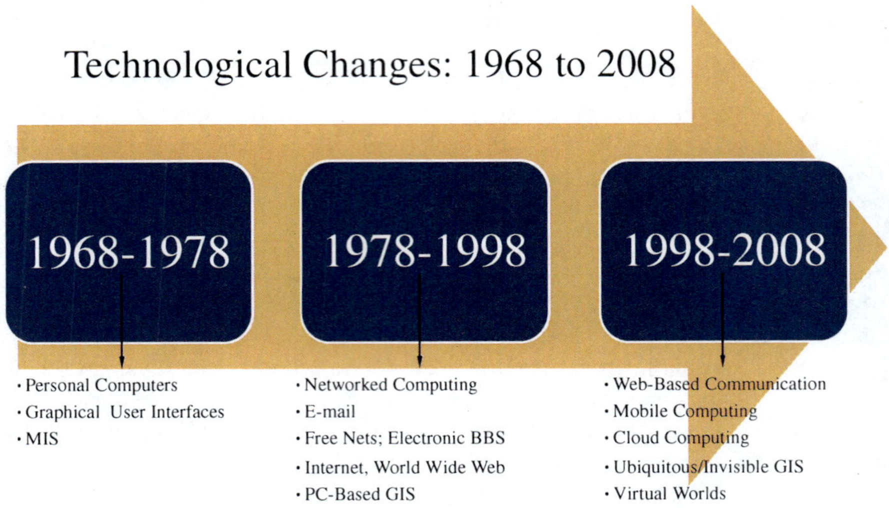

Fig. 2.1 Technological changes 1968–2008

the USA in the 1980s were the science agencies (NASA for example), while the main users were the utility companies and forestry sectors.

The notion of using GIS to support urban planning and management issues gradually emerged as the adoption and use of GIS among local governments began to spread. Huxhold (1991) was one of the earliest research-oriented practitioners who appreciated the value and role that information played within urban decision making processes. As he developed and managed spatial technologies for the City of Milwaukee over a period of 15 years, he learned a lot about the potential uses of GIS to address urban problems. Some of the highlights of his work include the development of maps to better understand the distribution of liquor licenses through the city and the identification of the worst landlords in the City "the dirty dozen".

Huxhold, like the authors of the URBIS study[4] (Kraemer, King, Dunkle & Lane, 1989) discovered that in government, managers exerted considerable influence in determining how data and information was collected, analyzed, and used. Thus, Huxhold concluded that the data and information needs for an organization focused on policy formulation were likely be substantially different from the needs of an organization intent on providing efficient services. For example, policy makers are likely to require integrated information that links the work of different departments; information that allows them to answer complex questions. Whereas, an organization primarily focused on efficient service delivery, is more likely be concerned about the accuracy of information it maintains, as well as the cost of its maintenance.

Huxhold discovered that while the potential of technology (GIS) use at the policy or decision-making level is high, the actual use of technology (GIS) in city government, as exemplified by the City of Milwaukee occurred through the automation of routine operations, that is, the computerization of tasks previously performed manually.

In the first two decades after GIS developed as a separate field and discipline, the dominant adopters and users of GIS were government agencies, research centers and universities. Universities using GIS developed new analytical tools, better database management procedures, addressed problems with spatial geo-referencing, error modeling, data quality management, and feature generalization (e.g., Poiker, 1976).

During these two decades, ordinary citizens and their more organized surrogates (community-based organizations), energized by earlier anti-war and civil rights movements, gradually redirected their energies to addressing environmental and public health concerns. As citizens mobilized to address issues of concern to the public at large, the government enacted laws to ensure that any project or activity receiving federal funds would specifically examine social, economic, and environmental impacts.[5] In most of these situations, the analyses were run by experts politically allied with the community cause.

2.2.3 Being Connected – Anywhere, Anytime (1988–1998)

Although the idea of a global network to electronically link all computer users had been around for a while, the electronic superhighway came closer to being a reality in the early 1990s. The digital revolution of the early 1990s was full of hype and unbridled hope. In particular, proponents of the information technology paradigm argued that the world was being transformed by emerging technologies for the better. Negroponte's (1995) book *Being Digital*, a *New York Times* best seller, offered an optimistic vision of the future transformed by the power of digital information.

> Early in the next millennium, your phone won't ring indiscriminately; it will receive, sort, and perhaps respond to your incoming calls like a well-trained English butler. Mass media will be redefined by systems for transmitting and receiving personalized information and entertainment. Schools will change to become more like museums and playgrounds for children to assemble ideas and socialize with other children from all over the world. The digital planet will look and feel like the head of a pin. (Negroponte, 1995, p. 6)

At the same time, dystopian views about the dangers of high technologies were also expressed with equal aplomb. Many people invoked Orwell's classic novel *1984* warning us about the horror of surveillant media technologies. Others like Clifford Stoll in his book *Silicon Snake Oil: Second Thoughts on the Information Highway*, observed:

> The key ingredient of their silicon snake oil is a technocratic belief that computers and networks will make a better society, [that] access to information, better communications, and electronic programs can cure social problems. I don't believe them. There are no simple technological solutions to social problems (Stoll, 1995, p. 50).

An entire genre of literature about how to cope, that is to say, how to survive life in a permanently networked society began to emerge during this time (Shenk, 1997). Critics also argued that the digital revolution contributed to the isolation and marginalization of individuals and communities. This is because access to the internet is not evenly distributed. The term of choice to describe disparities in access is the "digital divide". This divide has been identified both in terms of lack of access to the basic tools (Norris, 2001) and as well as skills to use the technology (Mossberger et al., 2003). In earlier days, connectivity (putting the fiber-optic network in place) was conceptualized as a barrier to access,[6] although with the development of wireless broad band networks, the problems of connecting remotely situated end users has largely been overcome in the United States. Yet, the data from the Pew Internet and American Life Project reported by Mossberger et al. (2008) suggests that many Americans (approximately 27% according to the report) are not connected to the Internet for one reason or another and cannot participate in our brave new digital world.

These revelations broadened discussions about access to consider the social and institutional contexts that can either provide or impede access to information. Likewise, the ability of the individual or group to be able to interpret and thereby use the information they have managed to obtain (sometimes discussed under the rubrics of digital literacy or digital citizenship) are also topics that concern practitioners and policy makers, who want to promote easy access to planning-related information. Presently, discussions about access includes topics such as freedom of information, individual privacy rights, the commodification of information, data quality, data sharing standards, spatial literacy, and the role of intermediaries (e.g., non-governmental organizations) in assisting the public to gain access to information (Craglia & Masser, 2003; Ramasubramanian, 2007).

The GIS discipline and profession benefited from the rapid growth of personal computers, improved analysis software, and the availability of census data. As GIS applications grew and expanded, complexities associated with their adoption and use became apparent. In response to the need for basic and applied research about GIS development, a consortium of universities established the National Center for Geographic Information and Analysis (NCGIA)[7] in 1989. The center began organizing its research efforts through a series of research initiatives. Predictably, the earliest research initiatives addressed technical issues (accuracy, data quality, and spatial decision support) but the consortium gradually began considering a broader set of issues including the sharing of geographic information (Initiative 9); user interfaces for GIS (Initiative 13), and collaborative spatial decision-making (Initiative 17).

NCGIA Initiative 19 sought to investigate the "social implications of how people, space, and the environment are represented in GIS". The "GIS & Society" initiative grew out of a 1993 workshop in Friday Harbor, Washington. A community of scholars began investigating the consequences of rapid adoption and use of high technologies and their impacts on the lives of ordinary citizens. John Pickles, in a seminal 1995 book, discussed the social implications of GIS charging the GIS community to pay close attention to the built-in biases that shaped GIS adoption and use. Wegener and Masser (1996) proposed alternative scenarios of the world in the year 2015 depending on the models of GIS technology diffusion and technology policies that were adopted in the 1990s. Their fears about the dominance of the market (the market scenario) have largely not come to pass, at least in the United States. It is reassuring to note that a lot of public information still remains accessible to the public, in part because of the assistance of the market.[8]

At the same time, electronic community networks grew rapidly, helping to link ordinary citizens. IT practitioners frequently use the term "community networks" to refer to electronic networks designed to foster community and advance social goals such as building community awareness, encouraging involvement in local decision making, or developing economic opportunities in disadvantaged communities (Schuler, 1997). Howard Rheingold (1993), based on evidence gathered from one of the world's earliest electronic communities – the Whole Earth 'Lectronic Link (the WELL) observed that computer-mediated communications provided *social network capital* (the capacity to meet others with similar interests, the readymade community), *knowledge capital* (the capacity to get on the network and ask for help on a range of subjects from a gathered community with diverse experience and expertise), and a sense of *communion* (being supported emotionally by an invisible community).

Anne Beamish, as early as 1995, organized these non-profit community networks into four broad categories:

- Freenets (loosely organized, community-based, volunteer-managed electronic network services. Freenets provide local and global information sharing and discussion at no charge to the Freenet user or patron).
- Bulletin Board Systems (typically stand-alone systems which focus on local information and discussion in a particular neighborhood or part of a neighborhood accessible only to those users who can connect to a particular computer by modem).
- Government sponsored networks (city-wide networks that are sponsored by state or local governments whose primary purpose is to make city records and municipal information available to all residents).
- Wired cities (a community or a city within which all kinds of electronic communications services are available to households and businesses or any experiment or project involving the provision of information and communication technology to households and businesses).

Schuler (1997) argued that network users would form an on-line community and behave as they would in a real world community, i.e., individuals and organizations would come and go as they pleased and participate to the intensity they desired; and in turn this participation would depend on their needs and interests as well as the openness, policies, politics, and spirit of the on-line community.

2.2.4 *Creating Empowered Netizens (1998–2008)*

Since 1998, the changes that have occurred in the world of computing and the development of new technologies have been mind-boggling. In 1994, the futurist John Naisbitt made a far-reaching pronouncement. He argued that the functional differences between the telephone, television, and personal computer would disappear (Naisbit, 1994). This phenomenon has largely come true. In 2008, the computer behaves like a television, a social networking portal, a mobile workspace, and when necessary, serves as a telephone with an optional video display. The costs of personal computers have declined further, and most people now use their mobile telephone to organize information, communicate via email, send pictures and video links to friends, listen to music, and find their way around town. Digitally empowered citizens are at the forefront of community organizing; social networking sites like MySpace and Facebook allow for multiple opportunities where online activities can spill over into the real world. Individuals using services like Twitter can provide news and information in real-time about on-going events at a pace comparable with, or better than conventional news media. In the next section, I will trace the evolution of Public Participation GIS (PPGIS) over this time period.

2.3 The Evolution of Public Participation GIS

Public Participation GIS is an awkward phrase that has come to encapsulate the intersection of community interests and the widespread adoption of GIS technology. As one reviews the social history of the field, it is interesting to note that the name choice PP+GIS emerged from the planning field[9] (Obermeyer, 1998). The early origins of PPGIS were focused on harnessing the capacities of GIS to serve community interests, while remaining cognizant of the potential limits of the technologies themselves. Even a recent exhaustive review of the subject (Sieber, 2006) failed to provide a clear definition of PPGIS, opting instead to characterize PPGIS as a field or a broad umbrella of practice activities, emerging from various disciplines and driven by disparate agendas. We can safely state that PPGIS is a term used to describe a range of participatory planning activities that are supported or enhanced by the use of digital tools such as GIS maps.

Despite ambiguity about its nomenclature (fortunately a distracting discussion that is limited to the academic enterprise), PPGIS adoption, or in other words, the use of GIS tools and techniques to solve a variety of community-oriented problems grew rapidly in the early 1990s. This spurt in GIS activity at the grassroots can be correlated with wider technology growth trends of the 1990s. In the United States, this growth spurt was directly and indirectly supported by the investments made by the federal government in the areas of education, health care, business, commerce, and environmental management, and in community development.[10] For example, between 1995 and 2000, US Department of Commerce[11] funded over a hundred projects including demonstration projects,

Table 2.1 IT/GIS applications for low-income communities

Application	Function	Primary/ secondary sites
Community Assessment Database	PC-based community development data sets; property/ infrastructure inventory	Community agencies; community development corporations; universities
Community Internet Server	Free-net provision of e-mail, telnet, ftp, Internet access, etc., for citizen use	Homes; community-based organizations
Interactive Crime Response Network	Electronic network to coordinate crime monitoring and public safety planning	Homes; community policing offices; community agencies
Information Kiosks	Single or multi-purpose electronic kiosks for information dissemination/ communications	High volume public access points
Community Health Station	On-line information on health promotion/ prevention for self-diagnosis/ referral	Community-based organizations; health and social service agencies

community networking projects, and infrastructure development projects all designed to improve electronic telecommunications and showcase the advantages of connectivity.

One of the earliest descriptions of IT applications designed to serve "low income" communities came from Richard Krieg (1995). Although the "PPGIS" terminology was not used in his survey, many of the applications and functions listed are examples of community-oriented spatially referenced information systems. At the time of Krieg's survey, many providers and consumers of information strove to bridge the digital divide by providing free or low cost access to e-mail and the Internet. Other applications required users to be at particular physical locations to access services (e.g., the offices of community agencies, public libraries, and other high volume access points). An overview of some of these applications is provided in Table 2.1.

While technology (the hardware) was seen as a primary barrier to bridging the digital divide, other barriers such as software, technical and literacy skills, as well as access to data were beginning to be recognized. The federal government's investment in technology access projects during this period cannot be underestimated. At the same time, community-based organizations in the United States were being challenged to take on additional service provision and advocacy responsibilities with limited resources. Creative community-based organizations were quick to explore the potential of emerging technologies to help achieve organizational goals. In some instances the traditional funders of community-based organizing and development provided funding for technology-related projects, while industry provided hardware and software donations.

By 1995, the US Department of Housing and Urban Development was requiring community-based organizations to develop applications to demonstrate community need in order to be eligible to receive block grant funding.[12] Community organizers discovered that by mapping census data and integrating it with additional information gathered from other city and county sources, they could begin to create a socio-spatial narrative that was more evocative to describe neighborhoods in need. Thus, the mid-1990s efforts tended to map misery (e.g., crime, socio-economic deprivation), with the goal of drawing precise geographic boundaries to target areas of greatest need. However, they spurred a culture of data-driven analysis of social issues that facilitated data gathering and data integration. Many of the nation's smaller cities received additional support for these efforts from philanthropic institutions[13] and research policy institutes.[14] The planning literature cites a plethora of small community-focused GIS activities during this time (e.g., Myers et al., 1995; Talen, 2000).

In 1997, an ESRI[15] publication, *Zeroing In: Geographic Information Systems at work in the community* (Mitchell, 1997) catalogued the use of GIS for a variety of social applications, including emergency dispatch, finding funding to build low-cost housing, tracking drug activity, and managing urban sprawl. Each example described a simple story, with readily identifiable and manageable problems and a structured set of solutions. A collection of case studies, including some that explicitly discuss PPGIS work in US community-based organizations, can be found in a compendium, *Community Participation and GIS*, edited by Craig, Harris and Weiner (2002).

In the nation's larger cities, comprehensive community building initiatives also encouraged data collection, integration and a managerial approach to social problem solving. Community-based organizations began providing access to real property and infrastructure inventories on stand-alone computers in order to better understand the dynamics of neighborhood change. Using an indicators-based approach, community groups were able to target physical interventions that were intended to address social problems (e.g., removing abandoned/boarded up houses to reduce risk of arson or drug crime). These systems eventually evolved into Neighborhood Early Warning Systems which were adopted in many cities such as Minneapolis, Chicago, Philadelphia and Los Angeles among others (Snow, Pettit & Turner, 2004).

Sawicki and Peterman (2002) using data from a 1998 national survey designed to assess the extent of PPGIS practice reported that a wide range of nonprofits, some affiliated with universities, as well as some government agencies were engaged in some kind of PPGIS activity. The 18 university affiliated projects identified in the Sawicki/Peterman study included centers that provided mapping and technical assistance services such as the East St. Louis Action Research Project[16] (ESLARP), and Neighborhood Knowledge Los Angeles[17] (NKLA). By this time, the web had matured to support internet-based data delivery. Government agencies were just beginning to get involved in data provision and dissemination via the web, with the lead being taken by federal departments such as the US Census Bureau, the US Department of Housing and Urban Development, and the Environmental Protection Agency. These national initiatives had their counterparts at the state and local levels of government. Many cities launched data delivery services with support from local and regional partners. One such example is the

Boston Foundation-funded project called The Boston Indicators Project. As the project's tag line – measuring what we value – suggests, this on-going initiative "seeks to democratize access to information, foster informed public discourse, track progress on shared civic goals, and report on change in 10 sectors" (Boston Foundation, 2009).

2.4 Changes in PPGIS Use and Planning Practice

During the 15-year time frame that participatory planning using digital technologies have been in vogue, many of the benefits of digital connectivity and access to geo-spatial technologies are associated with social learning. PPGIS advocates have emphasized "jumping scale", as one of the benefits of using GIS to address the needs of marginalized communities. It is argued that by allowing end users to explore issues at different spatial and temporal scales, these users were likely to be freed from the limits of their particular marginalized positions. It was also argued that by changing the units of analysis, new understandings and new alliances could be established and brought to bear in the problem solving process. It must be noted that while Rheingold (1993) and others in the digital networking world emphasized individual-to-individual and individual-to-group connections, those within community organizing/development world focus on community-to-community connections (Agre & Schuler, 1997).

In 2007, the goals of individual learning, development and empowerment have largely been achieved because spatial technologies have been more seamlessly integrated within ubiquitous applications. At present, even naïve users can explore a highly context-sensitive, communicative visualization that can provide multimedia experiences in real time. If access to information can be equated to empowerment, then, "netizens" have multiple opportunities to access information that reflects different perspectives and viewpoints. Individuals can take advantage of 3D interactive visualization tools such Google Earth™ to add data and information about particular issues without having access to complex GI technologies or software. These developments confirm and document another benefit emphasized by PPGIS advocates — that the technologies allow end users to participate in the production of knowledge, rather than remaining passive consumers of information (Gaventa, 1993; Ramasubramanian, 2004).

Individuals with minimal technical knowledge can now add ideas, comments, pictures, maps, and other kinds of data and information to enrich on-going conversations about a variety of social issues. For example, in an on-going fracas that pits community activists in Brooklyn, New York, against mega developers promoting a new stadium and high-rise residential complex in downtown Brooklyn, community activists have used interactive 3D visualization tools to "show" how the scale of the proposed development conflicted with the existing character of the neighborhood.[18]

The results are more mixed when the goals are neighborhood or community empowerment. Nonprofit organizations[19] now play an important role in facilitating PPGIS efforts. Local data providers include community-based service providers and advocacy groups. These organizations often create customized data sets that organize information relevant to a particular population subgroup (e.g., caregivers of young children) or by geographic boundaries that are more easily understood by ordinary citizens (e.g., neighborhood areas rather than census tracts). Community data centers[20] are also repositories of rich local and contextual knowledge. Community archives often include geo-referenced information that may not be available in official records. Examples of such local information include oral histories, drawings, sketches, photographs, as well as video and film clips.

While there is clear evidence that the stakeholders who got involved on behalf of the neighborhood are transformed through and because of their activism, it is difficult to assess spillover effects. Specifically, it is not clear whether community members who were not directly involved feel a sense of empowerment although they may have been among the ultimate beneficiaries of citizen activism and community-based planning. Furthermore, in documenting and evaluating spillover effects, it is difficult to determine the unique contributions of geo-spatial technologies.

The results of technology adoption to address a range of social issues are clear – in advocacy and participatory planning work, GIS is now part of the organizing arsenal required to challenge "official" planning decisions and policies, often generating new data and information. These new forms of evidence have served well the quest to energize citizen activism at the neighborhood scale. Yet, the results are not as clear when we seek to understand the transformative and collective impacts of participatory projects that used GIS, perhaps because published narratives of Public Participation GIS (PPGIS) adoption and use often focus on the particular case (e.g., Craig et al., 2002), placing little or no emphasis on the larger planning frameworks that govern technology adoption and use.

Furthermore, there are many activities that carry the PPGIS label causing great confusion among practitioners about what constitutes a PPGIS activity. While there are many researchers developing tools and methods to support PPGIS work (e.g., Lowry et al., 2009), there is no clarity about what ideal PPGIS activity should look like. In Chap. 3, I take on this challenge.

Notes

1. The adoption and diffusion of technological innovations into society's mainstream typically takes the form of an S-curve (Rogers, 1995), when there is enough of a critical mass of people (users, if you will) who have adopted a particular technology.
2. For a good historical overview of GIS, read Tim Foresman's (1997) book.
3. See, The GIS History Project archives at: www.ncgia.buffalo.edu/gishist/.
4. Ken Kraemer and John King of the Public Policy Research Organization at the University of California at Irvine, along with other colleagues, conducted detailed case studies of local

governments in seven cities in order to understand the complexities of adoption of information technologies. The Urban Information Systems (URBIS) project, begun in 1973 was one of the most comprehensive case studies of local governments' use of information technologies. The study sought to identify those policies (related to the management of information systems) that were most conducive to the adoption and utilization of information technologies within the organization, treating the organization as a comprehensive political and administrative system.

5. For example, The National Environmental Policy Act of 1969 (NEPA) is federal government legislation that requires a thorough analysis of the impacts of a project or activity receiving federal funds and requires consideration of social, economic, and environmental concerns.
6. The *universal service* provisions of the United States communications laws were originally intended to provide affordable local telephone service. The Telecommunications Act of 1996 expanded these provisions to include access to advanced telecommunications services at discounted rates to all communities, with a special focus on elementary and secondary schools, libraries, health care providers, as well as rural or isolated populations. For more information, see the Federal Communications Commission's website: http://www.fcc.gov/wcb/tapd/universal_service/
7. The National Center for Geographic Information and Analysis (NCGIA), Maine, the center is a multi-institution, multi-disciplinary research consortium dedicated to basic research and education in GIScience and related technologies. University of Buffalo: www.ncgia.buffalo.edu; University of Maine: www.ncgia.main.edu; University of California, Santa Barbara: www.ncgia.ucsb.edu
8. For instance, http://www.zillow.com/webtools/data-resources/
9. Obermeyer credited Dr. Xavier Lopez, then a student in Orono, Maine for suggesting this term; this was confirmed by Dr. Lopez through personal communication with Dr. Ramasubramanian in 2008.
10. Community development has been defined as a process "designed to create conditions of economic and social progress with the active participation of the whole community and with the fullest possible reliance on the community's initiative" (Rothman, 1974, cf. Levine & Perkins, 1997, p. 336).
11. The Telecommunications and Information Infrastructure Assistance Program (TIIAP), one of the programs of the National Telecommunications and Information Administration, is authorized by 47 USC-390-393A (1991) to provide resources to be used for the planning and construction of telecommunications networks for the provision of educational, cultural, health care, public information, public safety or other social services. It morphed into the Technology Opportunities Program (www.ntia.doc.gov/top/)
12. The Community Development Block Grant (CDBG) Program is among one of the oldest programs of the United States Department of Housing and Urban Development (HUD). It is a very flexible program that provides annual grants for a wide variety of activities related to physical planning including property acquisition, demolition, rehabilitation, construction of buildings, and economic development activities. CDBG funding mandates a high degree of citizen participation and an additional obligation that no less than 70% of CDBG funds are used for activities that benefit low-and moderate income persons. Additional information about the program is available at: http://www.hud.gov/offices/cpd/communitydevelopment/programs/index.cfm
13. Annie E. Casey Foundation's KIDS COUNT initiative is a national and state-by-state effort to track the status of children in the United States. The first national KIDS COUNT data book was published in 1990. For more information, see http://datacenter.kidscount.org
14. The National Neighborhood Indicators Partnership is led by The Urban Institute. The project began in 1996, funded by the Annie E. Casey Foundation and the Rockefeller Foundation. For more information see: http://www2.urban.org/nnip/index.htm
15. According to the company's website, ESRI was founded as Environmental Science Research Institute (ESRI) Inc., in 1969. A global company, headquartered in Redlands, California,

ESRI has been critical to the creation of a strong GIS user community in the United States. Additional information can be found at: www.esri.com

16. East St. Louis Action Research Project www.eslarp.uiuc.edu
17. Neighborhood Knowledge Los Angeles. nkla.ucla.edu was created in 1998 with a total project cost of over US $1 million with support from multiple sources with over half the support coming from the Technology Opportunities Program of the US Department of Commerce.
18. Barkey (2006) and Keegan (2006).
19. Neighborhood Data Center, a program of the NonProfit Center of Milwaukee, Inc., http://www.nonprofitcentermilwaukee.org/datacenter
20. Greater New Orleans Community Data Center, http://www.gnocdc.org/

Chapter 3
Dilemmas in Contemporary Planning

3.1 The Need for Frameworks

Definitions and frameworks are essential for the progress of systematic inquiry. They create the common ground that is necessary to engage in a meaningful conversation. While definitions can sometimes create intense disagreement and debate, they also help to forge consensus and advance scholarship (Wright et al., 1997). In the last chapter, I reviewed the literature on the evolution of Public Participation GIS (PPGIS) research and practice, commenting on the fact that there is still a lack of agreement about the term. Is PPGIS a set of tools? Is it a way of thinking about doing GIS work in communities? In what ways is PPGIS different from conventional GIS? Can PPGIS include more than GIS, for example, can we discuss it under the umbrella of Public Participation and Information Technologies[1] (PP-IT)?

To get us started on this inquiry, let me put forward a working definition that articulates both necessary and sufficient conditions for a particular activity, project, or program to be recognized as a PPGIS initiative. I stated that PPGIS activities are participatory planning initiatives supported by the use of digital technologies. This definition clarifies my world view – that digital technologies must be deployed in the service of a participatory planning agenda, not the other way around. Specifically, I propose that an ideal PPGIS/PPIT activity should:

1. develop the capacity of the participants to organize, analyze, and discuss planning concepts to the level required by the particular endeavor they are involved;
2. engage participants in every aspect of the planning process, that is, in the framing the project goals, the methods that are selected to examine and investigate these goals, in project implementation, and assessment;
3. develop techniques to carefully incorporate participants' views and participant-generated data into formal planning processes; and,
4. provide clear and transparent strategies for data generated from the project to be available to the participants.

L. Ramasubramanian, *Geographic Information Science and Public Participation*, Advances in Geographic Information Science, DOI 10.1007/978-3-540-75401-5_3,

This definition establishes a set of criteria that must be met by any PPGIS/PP-IT project. In the next section, I discuss how digital technologies are likely to impact and influence participatory planning processes. The discussion that follows is focused on the dynamic of involving people in the decision-making process, although it should be kept in mind, that there are overarching goals (e.g., build a new road, or create new affordable housing) that any planning activity seeks to achieve.

A typical participatory planning process considers:

1. Public Participation Goals
2. Participants
3. Methods of Community Engagement
4. Process Design and Management
5. Digital Tools
6. Data and Information
7. Project Timeline
8. Outcomes and Evaluation

The eight elements anchor my vision of PPGIS-practice and are discussed individually in Sect. 3.2.

3.2 Framework Elements

3.2.1 Participation Goals

In the flurry to discuss the participatory activities that are supported by technologies, many PPGIS researchers do not discuss the goals that drive any planning endeavor. A discussion regarding the goals/purpose of public participation in the United States would be incomplete without consideration of the work of Sherry Arnstein. Her "Ladder of Citizen Participation" (1969) has defined how planners conceptualize citizen participation. Much attention is paid to the Arnstein ladder, an eight rung typology that culminates with citizen power manifested through partnership (citizens "negotiate and engage in trade-offs with traditional power holders"), delegated power, and citizen control (where "have-not citizens obtain the majority of decision-making seats or full managerial power").

For Arnstein, the main purpose for engaging in a participatory process was to redistribute power – to give voice to those excluded from political and economic processes. Much of her thinking was based on the management of the federal anti-poverty programs, US Department of Housing and Urban Development's Model Cities program. To the extent that these federal initiatives were largely a product of top-down thinking; a hastily crafted government response to the civil unrest that prevailed at the time, one could argue that her analyses are limited and biased. Although Arnstein herself pointed out the many limitations of the typology, observing that the rungs of the ladder are simplistic abstractions of a more complex field of individuals, groups, and interests, subsequent scholars have tended to be strongly wedded to the static, unidirectional metaphor of the ladder. Furthermore, Arnstein's ladder is quite dated (40 years old in 2009) and by framing citizen control of government-led decision making as the only pathway to political power, she ignores the influence and contribution of other influential sectors in shaping American democracy.

Arnstein's ladder is a useful starting point in the discussion of the purpose of citizen participation. Other writers, including Wiedermann and Femers (1993) have also examined the issue of citizen participation goals, to explain why government agencies engage in citizen participation activities, creating incremental levels of involvement in different aspects of a formal planning process, ranging from education that has little or no impact on decision making to public participation in the final decision-making process.

In their meta-domain matrix that links public and participation, Schlossberg and Shuford (2005, p. 22), draw heavily from the Arnstein ladder and articulate the following goals – information, education, consultation, issue definition, joint planning, consensus, partnership and citizen control. Although Arnstein categorizes information and consultation as tokenism, she points out that "informing citizens of their rights and responsibilities can be an important first step towards legitimate public participation". Thus if a project/activity is limited to uni-directional information provision, then, it cannot be deemed a participatory planning activity.

The International Association for Public Participation (IAP2) takes a more pragmatic approach, linking the goals of public participation with increasing public impact on decision-making – their spectrum of public involvement goals begins with *information sharing*, ensuring that it is balanced and objective, and for the purpose of helping participants gain a better understanding of the problems, alternatives, and/or solutions; *consultation*, in order to obtain public feedback, *involvement*, to work directly with the public to ensure that public concerns and aspirations are consistently understood and considered, *collaboration* – acts of partnership in every aspect of the decision-making process, and *empowerment* – to place the final decision making in the hands of the public (Fig. 3.1).

In my view, the purpose or the overarching goal of any PPGIS endeavor is to enable the development of a critical consciousness "consceintizacão" (Freire, 1970). Critical consciousness balances active engagement within a problem-solving process with a reflective analysis of the process itself and the resulting outcomes. Because PPGIS implies the use of digital technologies within the participatory planning process, the technologies can be used in creative ways as part of the problem solving process and the reflection process.

Much has been written about critical consciousness, but for the moment, consider that the short term goals of a participatory GIS endeavor are to engage the creative capabilities of the participants in an analysis of their own circumstances, beginning with their experiential knowledge and gradually integrating this knowledge within larger knowledge structures in order to foster a dialogue and a conversation about the most serious concerns expressed by the participants. In community organizing,

IAP2 Public Participation Spectrum

Developed by the International Association for Public Participation

Increasing level of public inpact →

INFORM	CONSULT	INVOLVE	COLLABORATE	EMPOWER
Public Participation Goal:	Public Participation Goal:	Public Participation Goal:	Public Participation Goal:	Public Participation Goal:
To provide the public with balanced and objective information to assist them in understanding the problem, alternatives, opportunities and/or solutions.	To obtain public feedback on analysis, alternatives and/or decisions.	To work directly with the public throughout the process to ensure that public concerns and aspirations are consistently understood and considered.	To partner with the public in each aspect of the decision including the development of alternatives and the identification of the preferred solution.	To place final decision-making in the hands of the public.
Promise to the Public:	Promise to the Public:	Promise to the Public:	Promise to the Public:	Promise to the Public:
We will keep you informed.	We will keep you informed, listen to and acknowledge concerns and aspirations, and provide feedback on how public input influenced the decision.	We will work with you to ensure that your concerns and aspirations are directly reflected in the alternatives developed and provide feedback on how public input influenced the decision.	We will look to you for direct advice and innovation in formulating solutions and incorporate your advice and recommendations into the decisions to the maximum extent possible.	We will implement what you decide.
Example techniques to consider:	Example techniques to consider:	Example techniques to consider:	Example techniques to consider:	Example techniques to consider:
Fact sheets Web Sites Open houses	Public comment Focus groups Surveys Public meetings	Public comment Focus groups Surveys Public meetings	Citizen Advisory Committees Consensus-building Participatory decision-making	Citizen juries Ballots Delegated decisions

Fig. 3.1

where Freire's ideas have found favor, Rivera and Erlich (1992, p. 16) observe that (within the Freirian model):

> Organizers and communities [must] view each other as subjects, rather than objects, as learners, and as equals. The process of *conscientization* may be visualized as a double spiraling helix where both the organizer and the community learn from each other, the problems at hand and the strategies and tactics employed. Both parties become stronger actors because their learning is mutual, supportive, and liberating.

The metaphor of a double spiraling helix, a dynamic process of co-generative learning between initiators of participatory activities and the participants, facilitated by respectful dialogue is a theme that echoes through communicative action theory (Habermas, 1987), action science (Argyris et al., 1985), and transactive planning (Friedmann, 1992).

Some activist-planners like Angotti (2008) argue that the idea of consensus planning is a myth, and condemn the idea of participatory planning itself – labeling it a smokescreen designed to obscure the real issues. According to Angotti, participation is nothing more than "sitting silently at a public hearing or attending scores of meetings that have no significant role in making decisions that matter" (p. 29). Although he offers us no guidance about how to fix these problems, Angotti's comments make it clear that we must carefully consider participation methods and techniques, because the instrumentality of participatory planning often shapes our perceptions about the process and influences the outcomes as well.

3.2.2 Participants

To state the obvious, participants are central to any participatory planning activity. The academic literature, particularly that literature that emerges from geography has extensively examined the word "public" and the word "participation", in part because of the original framing of the phrase PPGIS (public + participation + GIS). Schlossberg and Shuford (2005) point out that "for PPGIS, the public can range from every resident in a neighborhood engaged with community asset mapping, to every US citizen[2] interested in viewing census data online". Based on a review of different framings and conceptualizations of the term "public", they offer a delineation of the word that includes as categories – decision makers, implementers, affected individuals, interested observers, and random public. They suggest that decision makers constitute a "simple public" in that they are well defined and few in number, making them easy to engage as a group. At the other end of the spectrum, the "random public" are a complex group, because of group size, heterogeneity, and spatially distributed (although this not explicitly stated by the authors). Other scholars like Reitbergen-McCracken and Narayan-Parker (1998) remind us that in the process of categorizing multiple publics, it is essential to identify potential beneficiaries, and those adversely affected, particularly those from vulnerable population groups.

The process of identifying multiple publics is a useful academic exercise and particularly efficient for designers of participatory projects so that they can plan and manage their consultative processes ahead of time. However, this approach may have the unintended consequence of creating "tags" and imply that participants can only play a particular role within a particular participatory project. This cannot be further from the reality – most of us wear many hats – and we don't necessarily switch roles when we are involved in a participatory process.

Creighton (2005) points out that we've used multiple terms to describe participants, including terms like publics, audiences, and stakeholders. Despite its limitations, I prefer the simpler word – participants. In my definition, the public includes individuals, informal groups, formal organizations, representatives of agencies, volunteers, and others *who have chosen* to participate in a planning endeavor. Undue emphasis on the term "public" is a distraction from a practical standpoint. It places more emphasis on the efforts of organizers and initiators of participatory planning processes to assemble a diverse group of participants rather than on understanding how the processes are managed, and the outcomes that can eventuate as a result.

In practical terms, a practitioner or an initiator of a PPGIS activity should begin work in their sphere of influence and expand outwards to gradually reach more individuals and groups. The practitioner should be aware that some potential participants (individuals or groups) may not be immediately identifiable and should make every effort to identify and engage them. But, the organizers should always ask themselves – why should a potential participant become involved? What are the barriers that prevent a potential participant form becoming involved? The answer to these questions can be better understood in the discussion that follows.

3.2.3 Methods of Engagement

There is an extremely long list of methods and techniques that can be used to facilitate participatory planning. Many of these techniques are catalogued in *The Public Participation Handbook*, Creighton (2005). Federal and state agencies have published books and communiqués about best practices in engaging the public.[3] Before the advent of digital technologies, the newspaper was the most important vehicle for information dissemination. Information can be gathered from participants in a variety of ways – from coffee klatsches, individual interviews, focus groups, and public hearings. In rural areas, rapid rural appraisal (RRA) and its more progressive counterpart, participatory rural appraisal (PRA) have been adopted to solicit and manage community responses. It's important to remember that the most successful community engagement techniques are customized to the particular situation and congruent with the cultural values and mores of the participant group.

In the United States, formal methods of public engagement are severely restricted, in part for legal reasons. Agencies cannot actively involve the participation of vulnerable populations[4] (e.g., youth, elderly, intellectually challenged, individuals receiving federal assistance) without following stringent protocols that govern engagement. Many of these laws were put in place to protect these population groups, but they have had the unintended consequence of formalizing and restricting engagement. Innes and Booher (2004) note a similar problem with open records and open meetings laws instituted in the United States to ensure participatory democracy. Open meeting laws, for example, require officials to publish meeting agendas ahead of time and not deviate from them. In their efforts to pay attention to due process, many agencies use Robert's Rules of Order (parliamentary procedure) to enforce order to the proceedings. Consequently many public meeting discussions are stilted and do not allow for the spontaneity that one would typically expect from a free-flowing discussion.

Architects and urban designers use a variety of visual techniques to solicit information and ideas from participants, although these are usually restricted to small groups. The work of architects is relevant in part because they use non-verbal techniques to communicate ideas. For example, The Center for Understanding the Built Environment (CUBE)[5] created a curriculum and materials to create "Box City". The materials allow children and adults to learn city planning principles, including creating opportunities for participants to talk about the future in a non-threatening environment.

Architects also rely on an intensive community engagement technique called a *charrette.*[6] The charrette is a consensus building strategy, usually conducted over 3–7 days, and can involve a large number of people who participate in different ways over the period of time when the charrette is conducted. The charrette is usually successful only when there is convergence around problem identification and when solutions are being discussed (Lennertz & Lutzenhiser, 2006). The charette's solution strategies tend to focus on physical interventions (sometimes called placemaking). The Oak Park project and the Common Ground Project discussed as case studies in Chaps. 6 and 7 explore how charettes can be adapted and expanded to integrate digital tools.

As digital technologies have become prevalent, they are being used to enhance conventional methods of community engagement. Now, faxes and email blasts are used to remind participants to attend meetings. Agencies and groups use RSS feeds[7] to remind subscribers about events and meetings of interest in a particular neighborhood or about a particular issue. Agencies use conventional mail surveys, phone interviews, and electronic surveys to solicit information and feedback about a variety of issues from the general public. These surveys are also disseminated in different languages. Agencies also use special methods (targeted outreach) to hear from disadvantaged population groups – most agencies have a designated public outreach coordinator, whose job it is to identify potential stakeholder groups and devise ways to reach them.

I propose that the following ten questions must be addressed in the process of selecting a particular community engagement method or group of technique.

1. Does the method of engagement identify overarching goals (the purpose of community engagement) explicitly to the participants?
2. Does the method of engagement outline a sequence of steps including intermediate milestones that must be achieved in the pursuit of overarching goals?
3. Does the method of engagement facilitate interactive communication and dialogue?
4. Does the method of engagement allow for participation of people without formal education or professional expertise?
5. Does the method of engagement allow for meaningful participation by vulnerable populations – young, old, frail, visually impaired, physically or intellectually challenged?

6. Does the method of engagement allow for the inclusion of graphical (non-textual) communication?
7. Does the method of engagement acknowledge the dominant realities (e.g., physical, social, political, cultural, and financial realities) of the participants?
8. Does the method of engagement offer participants new skills or strategies to describe the world around them?
9. Does the method of engagement foster discussion, analysis, and problem solving in teams/small groups?
10. Does the method of engagement provide learning pathways so that participants seeking additional information or insight have the ways and means to acquire it?

The answers to these questions will establish a clear link between the goals of public engagement with methods and techniques and provide structure and form, i.e., inform process design and management.

3.2.4 Process Design and Management

Public involvement experts will tell you the same thing that the text books tell you – preparation matters! In fact, it matters a great deal because it often determines the success or failure of a participatory planning initiative. The core team (members responsible for designing and implementing a participatory endeavor) must facilitate community input into the design of the process as early as feasible. Necessarily, these conversations will begin with community leaders and/or those previously involved in such activities. However, it must quickly expand to include a range of individual and institutional actors that are already active in that project context.

Process design requires the core team to plan the management of the field implementation. In thinking through and answering mundane questions such as: how many meetings, who's going to lead the meeting, what's the meeting format, where is the meeting going to be held, the core team is actually making significant decisions about the quality of the interactions and the outcomes that can result.

Typically, a participatory planning process must balance the need for large group meetings that focus on establishing community-wide agendas and visions with more intimate meeting formats that are suitable for detailed interactions about specific problems and issues. Both types of meeting formats must incorporate opportunities for interactive communication and feedback. In addition, a good process must provide points of entry for participants with different levels of interest and expertise to get involved. A good process must document and showcase major project milestones to ensure that those individuals, who did not have the opportunity to come to meetings can be informed. Much creative planning work actually takes place in intensive working group meetings with stakeholders. Thus a good process will pay particular attention to the process by which stakeholders are selected and invited to participate.

Process design must consider issues of staffing, including articulate spokespersons, technical (issue) experts, community (context) experts, meeting facilitators, graphics and visual communication specialists, writers, database/data managers, logistics coordinators, and individuals proficient in the use of print and digital media for mass communication. Staffing is directly related to resources and resource allocation. Good preparation will actively link available resources to ensure that all project activities are designed, implemented, and documented carefully. To state the obvious, it's better to do a few participatory activities and do them well.

Lastly, participatory process design and management is a craft. The core team members require practice and skills in addition to a deep enduring commitment to the spirit of participation. The day-to-day work of implementing a participatory planning process is extremely demanding; in particular, it requires the surrender of personal egos in order to achieve project goals.

3.2.5 Digital Tools

There are a wide range of digital tools[8] that can be used to support different aspects of participatory planning – some tools assist with the management of participatory processes discussed earlier. For example, participatory planning projects may benefit from the use of good content management systems[9] to archive and catalog data and information and help make them accessible to a wide variety of users.

In terms of using digital tools to facilitate community engagement, I propose that practitioners organize them into the following meta-categories after Mitchell (1999). In his discussion of communication alternatives and the consequences of the digital revolution, Mitchell (1999) proposed a simple 2×2 matrix to demonstrate how digital technologies can be adapted and integrated into day-to-day communication activities, considering that activities could be organized into synchronous or asynchronous modes of communication.

Following this logic, we can identify digital tools that facilitate communication in four ways:

(i) Physical place/synchronous mode (e.g., the use of digital tools like keypad polling in a community meeting);
(ii) Physical place/asynchronous mode (e.g., the use of a digital smart board that is placed in a community center allowing different individuals to review and make comments about a redevelopment plan using electronic "sticky notes");
(iii) Virtual place/synchronous mode (an online meeting where spatially disconnected participants view, listen, and respond to a single live presentation streamed via the web and participate in a discussion with attendees electronically in real time); and,
(iv) Virtual place/asynchronous mode (where individuals or groups are able to download customized content related to a project/plan to review at their own pace using their own computers/software).

In addition to this meta-categorization, the adoption of digital tools must be linked with the goals of participation, the methods of engagement, and the existing skills/resources available among the participants. If digital tools cannot be directly applied, then, the products that are generated from these tools can be used in more traditional ways – for example, printing a copy of a GIS-generated map and using paper copies for discussion at a community meeting.

GIS tools can be directly or indirectly deployed in all four modes of communication, although much of the PPGIS narratives and case studies are about the use of GIS tools in quadrant 1 (same place – same time). The benefits and constraints of using these tools in small group settings has been extensively discussed in the literature and are briefly summarized in Sect. 2.3.

A plethora of tools are now available to facilitate the creation of realistic 3 D representations (e.g., building photos)[10] and dynamic scenes (e.g., realistic simulations of traffic or pedestrian flows through recognizable urban settings).[11] These tools can be linked to conventional GIS map representations in order to further ease communication with naïve users.

Finally, there are a set of complex digital tools are better described as planning/decision support systems (PSS/DSS) (Brail & Klosterman, 2001). Planning support systems are generally cost-intensive assemblages that link the power of analytical models, the mapping and analysis capabilities of GIS, and the visual power of digital simulations. The most interesting developments in digitally enabled public participation are likely to emerge from the deployment of these systems (not tools) and they bear further investigation.

With ingenuity and some technical expertise, it is possible to create ad-hoc assemblages of tools and techniques as an alternative to the do-it-all planning support systems such as MetroQuest.[12] It is possible to use free software[13] to complete a range of common community engagement tasks, at least in the context of the United States, where data is widely available, and access to a personal computer and a cell phone are relatively wide spread. This leads us to consider the next element of the framework – data.

3.2.6 Data and Information

In the United States, much of the publically available data to examine urban issues, particularly data about socio-economic disparities comes from the Census Bureau. Over the years, many need-based programs or policies that investigate socio-economic disparities have used census data. Presently, socio-economic data is collected and assembled by a variety of governmental and non-governmental entities including research organizations, community groups, universities, and political parties.

The United States has made much of its geo-spatial data accessible to the public via a web portal www.geodata.gov, a one stop location with links to federal, state, and local geographic data. Data categories include administrative boundaries, atmosphere, business, demographics, health, transportation, and utilities to mention a few. Private entities like ESRI support a global network of geographic information users and data providers (www.geographynetwork.com).

In the United States, the focus is the availability of reliable micro-neighborhood data that creates meaningful data for local community-based problem solving. It is at this level of analysis that data disparities become apparent and consequential. For example, advocates for the homeless in large urban areas like New York City have to conduct a separate count to estimate the population of homeless individuals living on the streets. The Homeless Outreach Population Estimate (the HOPE count) is usually made possible with the help of concentrated volunteer efforts over a single night during the coldest months of the year. For smaller cities, these types of counts may be impossible to conduct on a systematic basis.

In recent years, advances in cell phone technology are being used to gather volunteered geographic information (VGI). These systems allows individuals to immediately record a wide variety of geo-tagged information and send it to a central location. In New York, the 311 system has enabled callers to upload images, videos and sound so that information and evidence that is often transitory (noise pollution or littering) can be documented (Rivera, 2007) although the data is only made available to the public at a highly aggregated scale. The aggregation avoids concerns about privacy and the fear of surveillance, but places limits on micro-neighborhood organizing efforts.

For conducting participatory planning efforts, the focus should be on data assembled by participants and reflect their needs and interests. Mike Barndt (2002) reminds us that the data collected should be "appropriate" to serve community needs. While the term appropriate is ill-defined, assembled data must be reliable (credible) in order to be useful to serve community needs. In many instances, the credibility of the data is associated with the individual or organization that creates and assembles the data for community use. While some advance planning can be done in assembling readily available socio-demographic data, much of the innovation will require new data collection that will take time.

References to data often focus on quantitative data. In participatory planning work, much of the community generated data is in the form of pictures, maps, drawings, check-lists, and sometimes verbal narratives. This data must be carefully compiled and archived so that it can be appropriately integrated in decision making.

3.2.7 Time

Participatory planning and capacity building processes as conceptualized in the beginning of this chapter takes time. In part, it takes time for a core planning team to get involved with community activities and not be regarded as complete outsiders. It is that investment of time that creates a *sense of trust* because people in marginalized communities often associate time commitment as a proxy for commitment to a particular socio-political cause, or to the community itself (Korten, 1986). The time that it takes for projects to be launched and implemented can take months, sometimes years. Archiving this process, along with milestones that record successes and failures is an important element of process design and management discussed earlier in this chapter.

Time must be closely correlated to project goals and the reasons for engaging in a participatory process, rather than the timelines imposed by external institutions such as funding agencies or universities. Insofar as one person cannot be expected to be available for project management through an extended time period, creating a distributed management structure that includes more than one project manager may be a necessary staffing strategy to ensure continuity.

3.2.8 Outcomes

I earlier argued that the overarching goal to create a participatory initiative is to enable the development of critical consciousness among participants. What is critical consciousness? Or in other words, how do participants who've become critical thinkers behave? This element is more extensively discussed in upcoming chapters (Chaps. 8 and 9) because it is so central to the issues we've raised thus far.

In brief, however, outcome can be thought of as short term gains and long term results. Distinguishing between short and long term is dependent on the situational context. In general, short term gains are measured at the end of the project period while long term gains are measured with an eye towards sustainability – the 3–5 year time period after a participatory initiative has formally concluded.

Another way to measure outcomes is by reflecting on the attitudes of participants: If[14]:

- the social, intellectual, and political capacity of the participants has improved;
- the participants become more articulate and effective advocates for their own and the community's interests;
- participants are more aware of the intricacies of urban governance and are better equipped to participate within these systems;
- there is increased community cohesion;
- there is willingness to participate, because there is increased trust in participatory processes and their outcomes; then, and only then can we confidently say that the investments of the participatory process have borne fruit.

Considering the linkages between goals and outcomes requires us to explore the similarities and differences between Citizen Science and PPGIS.

3.3 Citizen Science and PPGIS

Citizen science[15] describes the work of ordinary citizens who, while engaged in the pursuit of their own hobbies and interests make meaningful contributions to the work of scientists and researchers. However, this is not as simple as it sounds. A long standing and vigorous debate[16] persists between those scholars and researchers who are typically concerned with establishing and maintaining scientific rigor while others argue that science and scientific research should shake off its claims of neutrality and objectivity and engage more actively in solving immediate and pressing problems. Donald Schön described these tensions from the point of view of a practitioner thus:

> This dilemma of rigor or relevance arises more acutely in some areas of practice than in others. In the varied topography of professional practice, there is the high, hard ground where practitioners can make effective use of research-based theory and technique, and there is the swampy lowland where situations are confusing "messes" incapable of technical solution. The difficulty is that the problems of the high ground, however great their technical interest, are often relatively unimportant to clients or to the larger society, while in the swamp are the problems of greatest human concern. Shall the practitioner stay on the high, hard ground where he can practice rigorously, as he understands rigor, but where he is constrained to deal with the problems of relatively little social importance? Or shall he descend to the swamp where he can engage the most important and challenging problems if he's willing to forsake technical rigor? (Schön, 1983, p. 42).

Some researchers and scientists are unwilling or unable to work with non-scientists, either because of their own ideological biases preclude this option or because their research requires a high level of technical proficiency requiring skills not usually found among the general public. In recent years, researchers working to understand diverse natural or biological phenomena such as avian behavior, climate change, weather patterns, and spread of infectious diseases appear to benefit from working with non-experts who are simply interested in appreciating the phenomena as a hobby or encounter these phenomena as part of their day-to-day routines. For example, the Cornell University Lab of Ornithology[17] has a number of citizen science programs to collect data about bird populations, migration, nesting, breeding, and mating patterns of birds. In its most limited conceptualization, citizen science is about generating a reliable data stream for researchers.

Members of the public who get involved in these "scientific" activities can expand their own awareness and understanding about scientific issues. They are also likely to experience a sense of fulfillment/accomplishment because of their contributions to greater scientific and social goals. In addition, participants may gain access to new opportunities and experiences that benefit their own personal or professional growth.

Policymakers concerned about emerging threats consider citizen science methodologies promising in their efforts to cope with environmental and security threats. Citizen science approaches speed up information flow because observers can communicate information directly to researchers who can make sense of the data in a timely manner. The comparative advantage of engaging many individuals in the service of one goal can be summed up with two words, economy and efficiency. Concerns about accuracy and reliability do exist, but researchers engaged in citizen science work argue that providing training and guidance to participants already enthusiastic about the subject matter and committed to the scientific enterprise can overcome these obstacles.

The Open Street Map[18] is an example of a citizen science project that uses open source GIS tools and protocols to create a free, editable map of the world. The project has an affiliation with University College, London.

Planners working in poor communities have long emphasized the need to learn from the experience of the locals, and engage them in the development of planning and design solutions.[19] Likewise, since the early days when community-focused GIS work began, researchers engaged participants in data collection efforts. Historically, there were many reasons for this; the most obvious reason was the recognition by the researchers and research team that the local experience and expertise of community members was critical to the data gathering effort. Another reason was the lack of resources. Projects operating on limited or non-existent budgets relied on assistance and support from the community.

Citizen science projects have many similarities with participatory planning efforts that use GIS and other digital tools. Both approaches harness the capabilities of ordinary people to help solve problems. However, the majority of citizen science projects seem to focus on moving data, information and knowledge upstream – to researchers and policymakers while it is not clear how participation in these activities will provide immediate benefits to the participant beyond the self-satisfaction one gains from making a useful contribution to a larger societal goal.

Let us take a closer look at Park Scan,[20] a citizen science project launched in San Francisco and now being expanded to other cities. According to its creators, ParkScan is a "community-initiated, web-based reporting system that tracks maintenance conditions in San Francisco's parks and playgrounds. ParkScan has been helping to improve neighborhood parks and playgrounds since 2003". Using this web-based mapping tool, citizens and visitors can report on the physical and environmental conditions/characteristics of the city's parks. The project was developed by a San Francisco-based park advocacy group with the support of the city and is funded by two private foundations.

ParkScan has many positive attributes. Anyone can serve as an additional pair of eyes and ears to support the city's efforts to keep up its parks. In a time of tight budgets, the city staff probably finds it difficult to monitor all of the city's parks. At the same time, record keeping and monitoring help to measure progress and argue for additional resources. The data can be parsed to create different types of reports, e.g., reports that focus on particular issues (graffiti), park features or elements (trees or benches), according to political jurisdictions (districts), or according to the time taken to resolve complaints.

Yet, the entire effort seems to be focused on "helping" the city do its job better. Would it not be more useful to engage citizens in building a sense of community among park users? Would it not be more useful to encourage citizens to focus on programming activities in the parks? Perhaps the ParkScan website could be expanded to allow individual respondents to be able to speak and communicate with each other, rather than with the central office that receives and addresses their complaints. And, finally, we have to ask, where is the science in this effort – what scientific problem is being solved by using the time and efforts of volunteers?

3.4 Overview of Upcoming Chapters

The definition and the framework described extensively in Sect. 3.2 are based on a review of literature in many fields. The framework will be used to evaluate three case studies in upcoming Chaps. 5, 6, and 7. Before we move on to the case studies, I present to you, the results of a national survey of PP-GIS activities, conducted in 2008 in order to specifically focus on the complexities of doing participatory GIS work in the United States. Although the central argument of the book is that GIS activities are melding and merging with other digital technologies, the participants in the survey reflect a sizable community of users who are focused on what can be considered conventional GIS activities. Their experiences are of particular interest because they add an additional layer in framing the organizational contexts within which this work takes place.

Notes

1. CITIDEP (a research center on information technologies and participatory democracy) with headquarters in Lisbon, Portugal, runs several international conferences, including the 1st International Conference on Public Participation and Information Technologies Conference (ICPPIT99) held in Lisbon in 1999 followed up by a 2nd conference, ICPPIT03, held at MIT in 2003. Additional information about these conferences can be found at: www.citidep.pt
2. It is somewhat ironic that these authors are so keen to focus on "US citizens" when attempting to answer the question, "who is the public?" in discussions about PPGIS. Since when do you have to be a citizen of one country to be able to sample that country's census data, and if so, why should it be the case?
3. California Department of Transportation, http://www.dot.ca.gov/ser/vol1/sec1/ch3public/chap3.htm
4. Generally, these rules govern actions of government agencies that receive federal funding or universities. In a university context, these rules are enforced by the university's Institutional Review Board (IRB). Additional details about the IRB can be found at: www.hunter.cuny.edu/irb
5. Center for the Understanding the Built Environment http://www. cubekc.org/index.html
6. Lennertz and Lutzenhiser (2006) define a charrette as a multiple-day collaborative design and planning workshop held on-site and inclusive of all affected stakeholders.
7. RSS (Really Simple Syndication) feeds allows users with internet enabled devices to subscribe to content updates – typically, these updates refer to changes made to a website or a blog, but it can also be used to gather information about meetings and/or events related to particular topics.
8. The list of tools that can be used for facilitating participation grows every day. For this reason, it would be impossible to provide a list of tools, because it would become outdated very quickly.
9. Rhiza Labs has developed a proprietary tool called Catalog that helps end users explore, visualize, and analyze information from multiple sources. Learn more at: http://www.rhizalabs.com/products/catalog/overview/
10. Adobe's PhotoShop is a popular software program that allows photo editing and manipulation to create before-and-after scenes of urban environments. http://www.adobe.com/products/photoshop/family/
11. CommunityViz, Site Builder, Model Builder, Scenario 360 and Scenario 3D are all tools developed and distributed by Placeways, LLC (www.placeways.com).

12. MetroQuest is a proprietary set of tools that can be customized to support planning processes. Additional details available at www.metroquest.com
13. Google offers a variety of free services including Gmail (email), Google Groups (for creating mailing lists and discussion groups), SketchUp (3D renderings), GoogleDocs (ability to collaboratively write/edit/share documents, spreadsheets, and presentations), Blogger (a tool to develop and publish blogs), and Picasa (a tool to find and share photos).
14. With apologies to Rudyard Kipling.
15. Altan Irwin, in his 1995 book, *Citizen Science: A study of people, expertise, and sustainable development*. London: Routledge states that he chose the title Citizen Science because it was "pleasingly alliterative". He proposes that citizen science is a science that addresses the needs and concerns of citizens, one that is developed and enacted by citizens themselves. Thus, citizen science can also be about knowledge that is created outside of formal academic institutions and imbued with local, experiential evidence. Irwin does not privilege the knowledge created by citizen science over formal science.
16. There is a wealth of material about these classic debates. Alan Chalmers 1999 book, *What Is This Thing Called Science? An Assessment of the Nature and Status of Science and Its Methods*, published by Open University Press will provide a good introduction.
17. Cornell Ornithology Lab Citizen Science Projects http://www.birds.cornell.edu/ netcommunity/citsci/projects
18. The Open Street Map, http://www.openstreetmap.org
19. See John F.C. Turner's 1976 book, *Housing by People: Towards Autonomy in Building Environments*, London: Marion Boyars Publishers, and Christopher Alexander's 1985 book, *The Production of Houses*, London: Oxford University Press.
20. Park Scan http://www.parkscan.org

Chapter 4
PPGIS: State of the Practice

4.1 Introduction

I have previously observed that public involvement in planning is determined by particular social, political, and cultural contexts. Similarly, GIS adoption and implementation in planning is also influenced by a wide variety of contextual factors such as the attitudes of key decision makers towards the new technologies, availability of skilled personnel, and resource constraints (e.g., Masser & Onsrud, 1993; Campbell & Masser, 1995; Huxhold & Levinsohn, 1995; Obermeyer & Pinto, 2008). Even inter-departmental rivalries can influence how GIS adoption progresses within an organization (e.g., Kraemer et al., 1989).

Researchers use different strategies to examine these contextual variables and draw conclusions about their impacts and influence on the decision-making process. One such strategy is to use survey data to reflect on the state of the practice. Survey research has both advantages and limitations. Surveys, particularly those surveys distributed through the internet can reach large populations with relative ease. In this chapter, I will discuss the results of a national survey that I conducted in 2007–2008. The results provide a snapshot of the ways in which spatial technologies are currently being used to support and facilitate public participation. This survey effort builds upon earlier attempts to better understand PPGIS activities in the United States which are discussed briefly below.

4.2 The Sawicki/Peterman Survey (1996–1998)

In 1998, Sawicki and Peterman embarked on an ambitious project[1] – to produce a comprehensive inventory of PPGIS activities in the United States. Their work was informed by Craig's earlier analyses of the activities of non-profit data providers who were working to make public data accessible to community groups. Sawicki and Peterman quickly realized that the PPGIS field was growing rapidly and that there was little or no clarity about what constituted a PPGIS activity. At the same time, many individuals and organizations were avidly exploring the

L. Ramasubramanian, *Geographic Information Science and Public Participation*, Advances in Geographic Information Science, DOI 10.1007/978-3-540-75401-5_4,

capabilities of GI technologies in order to adapt these tools for use in community settings.

Sawicki's team cast a wide net to identify PPGIS groups, eventually identifying 65 organizations spread across 40 cities that were involved in some type of community-oriented GIS activity. This list included a wide range of nonprofits (30 organizations), some affiliated with universities, as well as some government agencies were engaged in some kind of PPGIS activity. The eighteen university affiliated projects identified in the Sawicki/Peterman study included centers that provided mapping and technical assistance services such as the East St. Louis Action Research Project[2] (ESLARP), and Neighborhood Knowledge Los Angeles[3] (NKLA).

One of the salient features of this survey is that it gets to one of the key issues embedded within the PPGIS framework I described in Chap. 2 – namely, the role of data and information. For in fact, the Sawicki/Peterman survey is really an inventory of data providers and data intermediaries; in the authors' words, "our goal is to assemble an inventory of organizations that contribute to public participation in community decision-making by providing local-area data to community groups" (p. 24). Other than an academic curiosity to understand the extent/spread of these activities, the researchers sought to draw some conclusions about the relationships between increased access to local-area data (i.e., sub-city/neighborhood level data) and community empowerment.

One of the more tangible findings from the survey was a confirmation that access to new data (particularly sensitive data, such as data about bank lending practices made available through the passage of the Home Mortgage Disclosure Act of 1975), coupled with GIS mapping capabilities did allow community groups to establish new ways of challenging systemic social barriers, as discussed in Chap. 1. This is an important finding. However, activists like Gale Cincotta were quick to point out that the legislation such as the Community Reinvestment Act were also essential to formalize and consolidate these gains,[4] a point reinforced in the case study in Chap. 5.

At the same time, a close reading of the analyses provided by the authors suggests that community-based organizations may have been ambivalent about the value of data and information. The authors found that some community groups did not really want to integrate generalized demographic data and information in their day-to-day decision making. From an organization's perspective, information about property ownership was considered interesting, perhaps because of its potential to create an organizing campaign against absentee landlords, but census data about neighborhood characteristics was not considered particularly relevant. From the discussion of the survey, one can conclude that many community-based organizations are likely to use data and simple spatial analyses to articulate the need for their continued existence to funding agencies, than to use data for the pursuit of community empowerment.

The Sawicki/Peterman survey served as an excellent starting point for my research, although, I was very cognizant that the field had changed rapidly in the intervening years.

The main focus of the 2008 PPGIS survey was to provide an updated understanding of the range of organizations that were engaged in PPGIS activity in the United States. The survey[5] consisted of three major components:

- a web search to identify a wide range of community-integrated GIS activities that could potentially be classified as PPGIS activity;
- a short electronic survey that was widely disseminated through professional and associational networks;
- a follow up telephone call/conversational interview with survey respondents who had indicated their availability and willingness to be interviewed.

4.3 Web Search

Like Sawicki and Peterman, I found that the definitional problem was a huge barrier in attempting to create an inventory of PPGIS activities. The web search identified a wide range of private, nonprofit, and government groups that were engaged in activities that are commonly associated with PPGIS work. These included unique commercial mapping sites, grassroots community organizations, on-line public agency data and mapping portals and university-community outreach centers.

In my attempt to organize the inventory, I reviewed earlier attempts at creating typologies. Leitner et al. (2002) had identified six ways in which GIS tools were being made available to community groups. These approaches included: (i) community-based (in house) GIS, based within a well-established community organization; (ii) university-community partnerships (limited engagement to achieve particular programmatic or project goals, such as engaging a GIS class in data collection/analysis activities at the behest of a particular community); (iii) GIS availability in public locations (e.g., in public libraries); (iv) Map rooms (a facility owned and managed by a city agency to generate customized maps to serve community needs); (v) Internet Map Servers (web portals that allow users to create and download maps and data via the Internet); and (vi) Neighborhood GIS centers (specialized community-based organizations that are focused on providing customized maps and analyses for community groups).

This typology was an excellent starting point to understand the different ways in which PPGIS activities can be supported. Yet, the Leitner typology does not include the various ways in which the private sector provides useful and relevant information to individuals and communities. In many instances, private developers of web services are pioneering citizen science projects discussed in the previous chapter. These portals allow citizens to add their own perspectives on the data they use and return it back to the wider community. In addition, the Leitner typology (because of when it was devised, in the late 1990s/early 2000s) does not fully incorporate the dramatic shift to web-based delivery of data and information by both private and public entities.

The results of the web search (see Table 4.1) were used to create a four part categorization that addresses the role played by the organizations. This simpler typology integrates the Leitner typology and the data intermediary typology (government agencies; university centers; quasi-autonomous non-governmental organizations; and non-profit organizations) devised by Sawicki/Peterman.

4.3.1 Community-Based PPGIS Facilitators

In this category are organizations that work directly with citizens to educate them about how to use geospatial data for the express purpose of improving their community. One such example is the work of the Recovery Action Learning Laboratory (RALLY),[6] based in New Orleans. The group that originally formed to assist with recovery efforts in post-Katrina New Orleans, has since focused on primary data collection, monitoring, and evaluation. The group uses the information to build local neighborhood capacity and initiate advocacy efforts. RALLY's work is supported through foundation grants and private donations.

4.3.2 University-Based PPGIS Facilitators

Both Sawicki/Peterman and Leitner et al. implicitly acknowledge the powerful role played by universities in providing intellectual and technical support, infrastructure and staff support to establish and sustain PPGIS projects. Many of these partnerships do not directly engage citizens in data collection and analysis. Rather, universities partner with local community groups that organize and redistribute the data or map products to community groups who request their services. The CUNY Mapping Service of the City University of New York assists a wide variety of groups in the design and development of community-oriented GIS activities, and produces maps and data for special projects on an as-needed basis. The service also supports the Open Accessible Space Information System (OASIS)[7] by maintaining a robust data inventory of open space and facilitating the creation of user-defined maps. Other examples of these partnerships include the London Air Quality Network[8] and Living Independently in Los Angeles.[9]

4.3.3 Data Providers

Data providers simply share a good amount of geospatial data with the general public. These providers can be community-based physical locations (e.g., the Data center in Milwaukee), or in virtual locations (e.g., dataplace.org) or exist in both real and virtual worlds.

Table 4.1 List of PPGIS applications

Application	URL	Category	Level	Features	How participatory?
Virtual Slaithwaite	http://www.ppgis.manchester.ac.uk/projects/slaithwaite	Land use planning	Beginner (Beg)	Citizens comment on a particular road, site, building, etc.	High: allows for comments
London Air Quality	http://www.londonair.org/uk/asp/virtualmaps.asp	Environmental	Advanced (Adv)	Create maps showing actual and future pollution conditions; data available for download	Mid: informs only; but good data for spatial decisionmaking
Walk Score	http://www.walkscore.com	Residential	Beg	Enter address, shows walkability "score" of your neighborhood and nearby goods / services	Low - informs only; little "hard" data for spatial decisionmaking
Chicago Crime Map	http://www.chicagocrime.org/map	Crime	Beg / Intermediate (Int)	Google base map, query crimes by location, time, etc.	Mid: informs only; but good data for spatial decisionmaking
Chicago Police CLEARmap	http://gis.chicagopolice.org	Crime	Int / Adv	Advanced GIS interface allows for creation of maps, tables, etc.	Mid: informs only; but good data for spatial decisionmaking
Open Street Map	http://www.openstreetmap.org	Cartography	Int / Adv	Free, unrestricted base map; user adds his own spatial data	High: public map creation; but requires skill
Primo Spot	http://www.primospot.com	Residential	Beg	Shows on-street parking availability for a given address; uses Google base map and interface	Mid: informs only; but user can suggest corrections
NYPRIG CMAP	http://cmap.nypirg.org	Partnership	N/A	Nonprofits partner with CMAP to produce maps to aid in their spatial decisionmaking and/or advocacy	Mid: GIS work is done by the expert (CMAP); but the nonprofits are getting a product to aid them

Table 4.1 (continued)

Application	URL	Category	Level	Features	How participatory?
NYC OASIS	http://www.oasisnyc.net	Partnership / comprehensive map	Beg / Int	Comprehensive city map; users can select a host of themes to create their own map	Mid: while map is simple; navigating and understanding the different layers requires skill and knowledge
Minneapolis Neighborhood Information System	http://www.cura.umn.edu/MNIS.php	Partnership	N/A	Similar to CMAP. Community Groups partner with MNIS, which is a university consortium, to receive GIS training.	High: appears that groups work with MNIS to learn GIS; not just a mapping service
Farmland Preservation and GIS	http://www.lic.wisc.edu/shapingdane/facilitation/agriculture/options/bulletin.htm	Partnership / service	Adv	University of Wisconsin researchers used public input to determine community concerns (farming), then created a GIS to designate Farmland priority zones)	Low: participatatory only at first stage in seeking community input; skilled researchers did the analysis
Portland, OR Metro Governement / City of Portland	http://www.metro-region.org/article.cfm?ArticleID=737	Partnership / service	N/A	Metro offers a range of GIS services and analysis (at a cost) for concerned citizens and groups. City of Portland maintains an impressive, web-based GIS mapping service.	Mid: impressive range of data presented, but to use offline requires an expensive partnership with the city or Metro
MetroQuest Software / GuelphQuest	http://guelph.ca/living.cfm?subCatID=1615&smocid=2193	Visioning software	Beg	Residents log onto city website and answer questions regarding their preferred growth patterns. Virtual scenario (i.e. map) is presented showing the effects of their choices	Mid: users are entering predefined responses and seeing their results; not clear what decisionmakers are doing with this information
Neighborhood Knowledge California (NKCA)	http://nkca.ucla.edu/	Partnership / community mapping service	Beg / Int	Multifunction service run by UCLA that allows a two-way exchange of data; users can download or upload data onto their own maps.	High: cf. two-way exchange of information
Neighborhood Knowledge Los Angeles (NKLA)	http://nkla.ucla.edu/	Partnership / community mapping service	Beg / Int	Similar to NKCA but confined to Los Angeles. More emphasis on the dissemination of official data such as tax and property records	High: suggests that users use their found information to influence land use decision
Living Independently in Los Angeles (LILA)	http://lila.ucla.edu/	Advocacy mapping service	Beg / Int	Web-based map of disability services in the LA area. Detailed information added by users in the disability community to stress relevant services and resources	High: relevant information is added based on neighborhood knowledge regarding the needs of the disabled.

Table 4.1 (continued)

Application	URL	Category	Level	Features	How Participatory?
Common Census Map	http://www.commoncensus.org/	Perception map	Beg	"Research?" experiment by an individual programmer. Users answer questions based on their local and regional perceptions; experimenter makes maps based on this data	High on user input, but output is defined by the experimenter
Crisis in Darfur - USHMM / Google Earth	http://www.ushmm.org/googleearth/	Advocacy / educational mapping	Beg / Int	Users download Google Earth layers that mark burned villages, show pictures and offer testimony	Low: users simply see genocide-related data; they are supposed to act on it by voicing their concern to government
Gentilly Neighborhood Mapping Center	http://icpd.dartmouth.edu/viewer.php	Disaster / advocacy mapping	Int	Map depicts rebuilding conditions in Gentilly, New Orleans. Users can register to send status updates for the condition of properties in their area	Mid: user-provided data, but GIS work is done by experts
NEDAP Financial Justice	http://www.nedap.org/programs/mapping.html	Advocacy mapping service	N/A	NEDAP partners with local community groups to produce economic-justice themed maps at little or no cost	Low: participatory in end-product only, as recipient groups are urged to petition government based on spatial data they are given
South African San Institute – Cultural Resources Asset Management	http://www.sanculture.org.za/body.htm	Cultural resource documentation	N/A	SASI is an NGO involved in documenting and teaching the indigenous SAN people how to record their local knowledge and foster cultural preservation. Mapping of local resources is an example	High: it appears that the group and the San work side-by-side so that the indigenous people become equipped to document their local knowledge
Philippine Association for Intercultural Development	http://www.iapad.org/pafid/about_pafid.html	Cultural resources / indigenous rights	N/A	PAFID partners with indigenous groups to help them create maps to establish boundary claims to ancestral lands among other activities centered on indigenous rights	High: as per the site, "The three-year project covering approximately 100 000 ha of ancestral domains will have three components: (i) participatory community mapping; (ii) ancestral domains management planning; and (iii) capacity-building. "

Many local governments have developed online GIS portals where citizens can look up information on their community or particular piece of property. Often totally web-based, data is usually for display purposes only and cannot be downloaded. One such example is the Chicago Police Department's[10] comprehensive website, where citizens can monitor and map crimes in the city, using official police records. However, citizens cannot add new information to the site to report crime events. Privacy concerns are also addressed; no exact addresses or names are available through the website.

4.3.4 Spontaneous PPGIS Activities

Our research identified a number of web-based mapping/analysis applications that are customized to meet the needs of a particular user community (e.g., bicyclists in New York or travelling salespeople). These web-based mapping applications use a freely available mapping platform like Google Earth and adapt it to display a specific attribute that is available at a user-selected location. For example, the website walkscore.com generates a "walkability" score for a user's selected neighborhood, based on the proximity of mapped services, recreation and transportation access.

These niche projects are the most interesting PPGIS applications by far. The tools and PPGIS projects are a result of innovative work by individuals who have integrated two or more disparate sources of data to create new web-based services. These applications, often called mashups[11] address specific community aspirations. Examples include Chicago Crime Map,[12] Trailhead Finder,[13] and HotSpotr,[14] and their number continues to grow. In some of these instances, the data is provided from existing public sources. For example, the Chicago Crime Map data comes from the Chicago Police department, although the Chicago Crime Map is not an official source of crime information. In other instances, data is willingly provided by individuals who participate in the initiative by entering information into an online database (e.g., where users enter data about wifi hotspots). There is great interest in the use of such volunteered geographic information to energize and foster PPGIS activities.[15]

These niche projects are characterized by their self-organizing capabilities and their ability to adapt to the needs of their community. They are very useful in providing end users with customized information about local neighborhoods, but there is no tangible evidence linking this kind of information access to the establishment of new participatory activities. On the other hand, these projects are also most likely to disappear once project-specific goals are achieved. For instance, the Chicago community-driven crime mapping website[16] discussed earlier has been shut down by the creators.

The experience that I had guiding a student-led PPGIS project highlights some of the complexities associated with niche projects. Hunter College students Jason Nu and Wallace Murray wanted to advocate for the establishment of additional bike parking facilities in strategic locations in New York City. The premise was that more people would ride their bikes to work if there were secure and easily accessible bike parking spots. User involvement was necessary to identify desired locations (actual sections of sidewalk where bike racks could be situated). Nu and Murray created a website that allowed users to post locations (points on a map) about desirable bike parking locations. When the project began, the students were concerned about lack of participation. They advertised their project and the website link on local blogs, particularly those blogs known to be bike-friendly. The project struck a nerve among the biking community in NYC, particularly in Manhattan and the website received a lot of hits! Over 800 locations for bike parking were suggested within the first 2 weeks. The data flood created technical and organizational problems for the students and they eventually had to shut the site down because they wanted to spend time analyzing the data that had been submitted. Unsurprisingly, many of the locations proposed by participants were proximate to other transportation modes, e.g., subway stations and bus stops. An interesting issue that came up through user comments was the need to engage owners of commercial buildings to provide secure bike storage for employees. The students found that established and regular bicyclists were more active on the site and that the needs and attitudes of casual bicyclists or non-bicyclists could not be gathered using the e-survey approach they had taken.

Once the project was complete, the students shared the results with the NYC Department of City Planning (DCP). Planners at NYC DCP were able to incorporate the students' research into the department's own strategic planning efforts. NYC has since expanded their commitment to bicycling in the city through a comprehensive strategy that includes the provision of bike racks near bus stops and in other strategic locations.

Niche projects can be set up quickly and phased out of existence; they have no need to perpetuate themselves. Once the City accepted the argument that bike racks were needed and began to plan for their placement, community advocacy about bike racks became less important. Niche projects are not sustainable in the long run unless they are transformed into commercial applications that can generate revenue through advertising or user contributions.

4.4 E-Survey

The web search and analysis of applications provided an overview of PPGIS activities (categorized along the 4 part typology, described above). It also generated a preliminary list of organizations that were targeted to receive the survey.

4.4.1 Survey Design/Distribution

The survey was designed and distributed using a free electronic survey development tool and hosting service.[17] The survey design and my dissemination strategies were reviewed by the Institutional Review Board (IRB)[18] at Hunter College. After receiving IRB approval, I distributed the web link to the e-survey to my personal and professional networks via email. Subsequently, a description of the research and links to the survey were posted on various listservs and message boards frequented by actual PPGIS practitioners (PPGIS.net, IAP2) or more general community development communication vehicles (COM-ORG listserv) during in the first quarter of 2008.

4.4.2 Survey Questions

The survey consisted for eight closed-ended questions, and two open-ended questions (see Appendix). Questions 1 and 2 document organizational characteristics and the type of GIS service/applications provided. Question 3 and 4 delineate discrepancies, if any, between the intended audience (primary target audience) and actual users. Question 5 addresses the issue of organizational sustainability (asking whether the services were free to use or fee-based services). Question 6 seeks to catalog the diversity of data sources used for community decision making. Questions 7 and 8 get to the heart of PPGIS work – asking whether end users can add their own data to create customized analyses and whether such data is reviewed or checked for accuracy.

The framing of questions 7 and 8 is a further clarification that I place great emphasis on two-way information way as an essential ingredient of PPGIS work. In Chap. 3, this issue is discussed under the framework element "Methods of Engagement". One would assume that most mapping services make some use of official data, for example a base map showing roads and town boundaries, major landmarks, or basic property data, yet I posit that participation can be significantly enhanced through the use of user-submitted data. In other words, a two-way flow of information can develop, as users add their own impressions and understanding of their community to an otherwise neutral map made up of lines and polygons. One example of the benefits of this approach is the Neighborhood Knowledge – Living Independently in Los Angeles disability awareness service, which relies on users to submit information regarding facilities and services they find helpful[19].

I sought to increase the response rate by minimizing the length of the survey, and ensuring anonymity. However, the final two questions provided respondents with an opportunity to leave their name, affiliation and contact information should they be interested in participating in follow up interviews.

Due to funding limitations, the e-survey was available only for a short time, through the months of February, March and April 2008. The following analysis represents a snapshot of responses as of April 1, 2008. However, the survey remained open through the month of April to collect responses from additional parties interested in our research. We collected between 115 and 126 valid responses (not everyone answered every question), although 258 individuals began the survey by electronically signing the informed consent form.

4.5 Analysis of Survey Results

4.5.1 Organizational Characteristics

In allowing survey respondents to characterize their organization, the question included eight separate descriptive phrases and a ninth "Other, please specify" option. However, respondents were only permitted to make a single choice – this required organizations to self-identify with the set of tasks that was most representative of their day-to-day work. Over one-third of respondents identified themselves as affiliated with a university in some way.

About 10% of respondents identified themselves as community-based, a category that included community development corporations (CDCs), community-based organizations (CBOs), or community-based data providers. Government data providers accounted for 13% of the respondents, while another 12% defined themselves as non-governmental organizations (NGOs), a category that included many of our international respondents. However, a full one-quarter of respondents selected "other". This category included respondents from private consulting firms, the United Nations and affiliated agencies, state and local government agencies, and members of the general public (Fig. 4.1).

In casting a wide net, the survey reached a wide variety of organizations that saw themselves as doing some sort of PPGIS work. However, an e-survey, just like any survey, has its limitations in terms of outreach. Established organizations with active staff responded to the survey. The majority of respondents were affiliated with academia or government. Since this survey is a sample of PPGIS practitioners, one

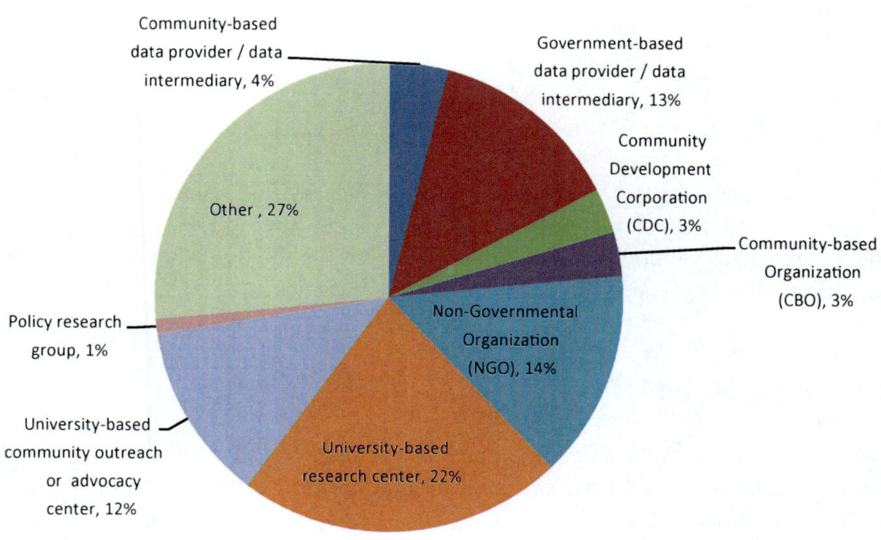

Fig. 4.1 Organizational profiles

cannot assume that this respondent profile truly captures the number and breadth of community-based PPGIS services.

The second question that focused on organizational goals, "which statement best describes your goals in offering GIS services or applications" allowed the respondent to select more than one choice. Close to two-thirds (61%) of respondents said that they hoped to encourage community involvement, while about one-third (36%) described their work as having more of an advocacy component. Information dissemination was also a popular organizational goal (61%). Other goals included more specific descriptions, such as "give people an idea of how issues and problems or opportunities look spatially and visually because statistics and [simple] graphs don't always do that well". Education was frequently cited as an organizational goal, among those who had selected the "Other" option, approximately (19% of respondents). This included education of "interested parties", "high school students", and "community members", and "policy makers" (Fig. 4.2).

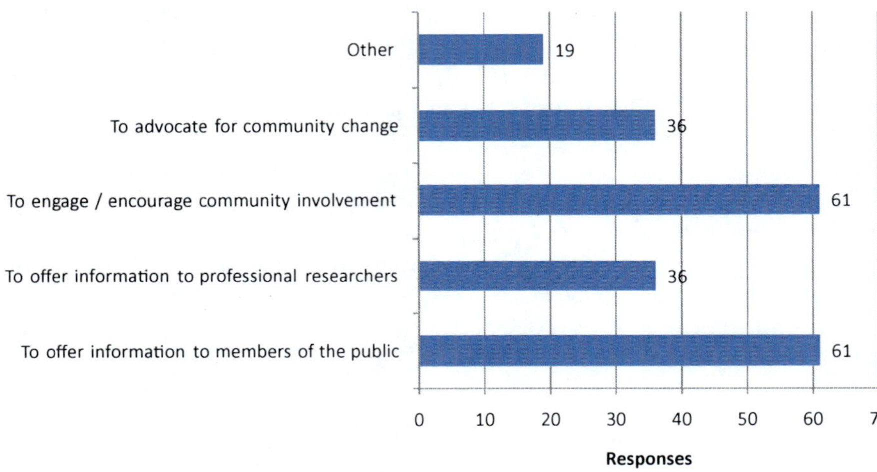

Fig. 4.2 Organizational goals

Drawing from the most prevalent responses to our goals questions, data provision (i.e. offering information) remains a top priority PPGIS activity. Yet, most PPGIS advocates would agree that data provision is a necessary but not a sufficient condition for an application to be considered PPGIS.

4.5.2 Intended and Actual Users of PPGIS Applications

Question 3 asked, "Who is your primary target audience?" in other words, who are you trying to reach? Unsurprisingly, a large majority of respondents selected the obvious answer "public" (45%). Other GIS application developers and service providers sought to reach professionals like community organizers (26%). The "other" category included "students/youth", "elected officials" (29%). Many people also selected "other" and wrote in a comment to state that they intended to serve "all of the above" which was not an option that I had offered in the framing of the question.

In response to the question, "who are the actual users of your services/applications", 42% of the respondents said that members of the public were the actual users. While 27% stated, "same as above", an overwhelming 72% stated that the actual users of their services were "professionals" (40%) or "community organizers/activists" (32%) (Fig. 4.3).

From these results, I conclude that PPGIS services and applications are being used to greater degree by a class of professional users – i.e., community organizers, researchers, journalists, policy makers, elected officials, and administrators of grant programs, rather than John and Jane Q. Public. It is also interesting to note that "youth/students" always stood out as a separate category, recorded under "other", rather than being included as part of the "public" category.

There is nothing wrong in these services being used by professional users – journalists and organizers are as much part of the community they live and work in,

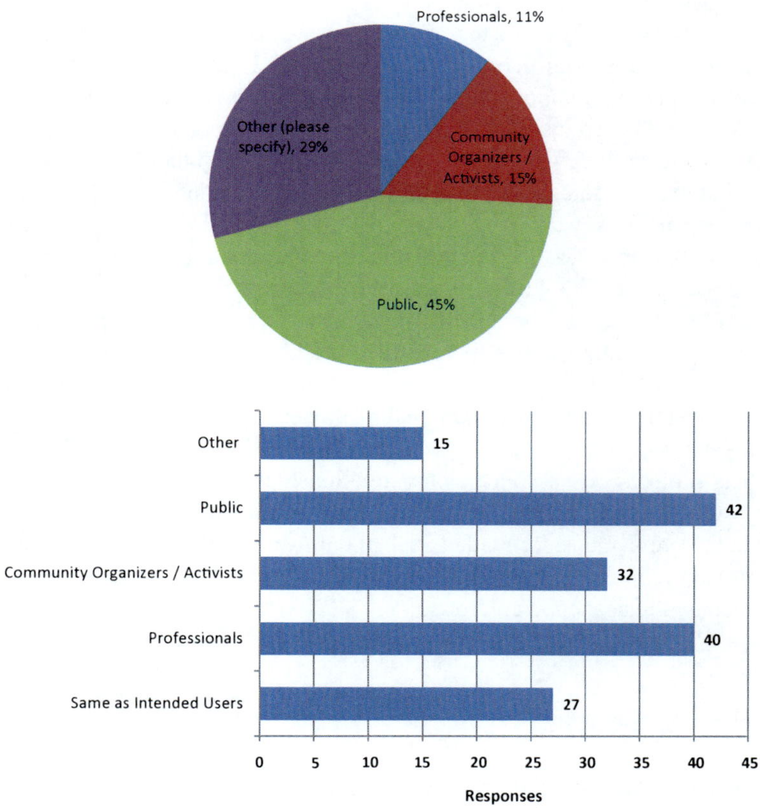

Fig. 4.3 Intended and actual users of PPGIS applications

and can be considered members of the public. However, the numbers do draw our attention to barriers – educational and technical barriers that continue to inhibit the widespread adoption and use of PPGIS tools and applications.

4.5.3 Organizational Sustainability

Many community-based GIS activities are conducted with limited budgets and with volunteer support. Still others depend on funding from philanthropic organizations and government sources to continue their work. Earlier studies (e.g., Ramasubramanian, 1998; Elwood, 2000) have found that local organizations use GIS capacity to support work that is internal to the organization (like grant-writing) in addition to using the tools' capabilities to directly support the mission of the organization. I wondered whether community-based organizations were using market-driven approaches to charge for services as a way to ensure organizational sustainability, but also as a way to manage demand (so as not to get overwhelmed by the number of service requests).

The survey results suggest that community-oriented PPGIS actors shun these market models (only 2% of respondents) said that they charged for services. Most respondents said that they made their data freely available, i.e., without charging for services and without requiring any user registration. I associated this finding with the rise of web-based mapping software. Coupled with a rise in Internet and broadband access, web-based mapping can make data quickly and easily available from a home PC. However, in reviewing the comments left in the "other" category (23%), I realized that many organizations and PPGIS service providers were grappling with this question, and most were considering creating password protected sites (limiting access) as well as investigating the potential for charging a fee to use their services.

The reluctance to charge for services is associated with the need to "democratize data", translated by organizations to mean free access to public data. The costs/burdens of adding value to readily available public data are being borne by the nonprofit sector (foundation grants and the like) or by universities (especially in university-community partnership projects). However, these two models – dependency on philanthropic subsidies and on the benevolence of university researchers/projects sets up dependency relationships and creates tensions related to data ownership and control – a theme that has not fully been explored in the academic literature. Considering that 38% of our respondents came from university-based research centers, PPGIS adoption and use appears to benefit from, and be constrained by the academic enterprise.

4.5.4 Data Sources

Official records still form the "backbone" of PPGIS work; almost 75% of respondents reported using some form of government-generated data related to demographics, land use, crime, and physical and social environment in their day-to-day work. Approximately 16% of respondents indicated that they harvested commercial records (information from yellow pages, real estate listings, advertisements) to create new analyses. Many respondents (52%) also indicated that they used direct observations – such as updating land use data received from the city files to correctly record vacant parcels or run-down properties to better reflect the realities experienced by their constituencies and to deliver more accurate/credible analyses. Community input (data provided by the public) was accepted as a reliable form of data input; over 41% of respondents said that their PPGIS systems allowed end users to submit their own data. In addition, approximately the same number of respondents (38%) reported that they verified user-submitted data for accuracy.

This is a wonderful finding, because the advantages of allowing for user input can easily be undermined if user-submitted data is incorrect or leads to faulty analyses. Checking the validity of user-submitted data, if done in such a way as not to undermine users' contributions, can be an important step in creating a more developed PPGIS system, one that uses objective and possibly subjective knowledge to paint an accurate portrayal of community perspectives. However, it appears that the respondents to our survey have not completely resolved the tensions associated allowing for user updates to "official" data. A case in point – while working with a community group to determine the age of buildings in a Brooklyn neighborhood, one of my graduate students found a string of buildings with the year 1940 as the year-built date. However, one of the buildings was a historic building. Working with a long-time community resident, she unearthed some old newspapers and retrieved information to indicate that the building was actually built at the turn of the last century (Brisbane, 2005). However, neither she nor the community group she worked with was able to convince the city to update its official records.[20] The city's approach to handling discrepancies in data is very different from the approach taken by Wikipedia – the dynamic encyclopedia project that allows users to edit information. The fears that entries will get hijacked by special interests have largely been unfounded; egregious offences have only occurred on a handful of topics/postings.

4.6 Interview Findings

A total of eight professionals in the PPGIS field were selected for a brief (10–15 min) telephone interview in March 2008. The sample was drawn from a universe of 51 survey respondents who had provided their contact information and had responded favorably to the question, "Are you open to having a longer conversation about the topics raised [in the survey]?"

Those interviewed included university professors, community-based groups and professional researchers. Respondents from outside the United States were not contacted for telephone interviews, because my research is focused on PPGIS activities in the United States.

This stage of research was designed to have PPGIS practitioners elaborate on concepts that were perhaps too broad for our e-survey. Primarily, the interviewer used this opportunity to speak with individuals about the extent to which the deployment of GI technologies influenced planning processes at the community level.

Understanding PPGIS outcomes proves to be a complex task. While all of the respondents were willing to discuss those "success stories" that they felt defined and validated their PPGIS work, many interviewees remarked that it was hard to precisely measure the effectiveness of their work. A university extension landscape architecture professor who works and trains participants explained the complexities in attempting to evaluate PPGIS activities. He observed that no two situations were alike, implying that a particular PPGIS application may work in one circumstance but fail in another. Context is everything, seemed to be the take home message, from these interviews. We will reflect more on the thorny issue of PPGIS evaluation in Chap. 8.

4.7 Next Steps

In practice, PPGIS work is a discontinuous series of steps; in which the first step is an attempt to make data/information/spatial analyses available to the public. Subsequently, the success of PPGIS activities depends largely on the management of the process – the ways in which individuals and groups are challenged to think critically about difficult problems. I am not trying to create divisiveness by distinguishing between "real" PPGIS and conventional mapping of social/physical assets and problems using GIS. Rather, the survey findings suggest that many individuals who are not working collaboratively with the community, still view themselves to be doing PPGIS work. The next three chapters take on a more in-depth look at case studies in three different contexts (the neighborhood, the city, and the region, respectively) to further articulate the dynamics of doing PPGIS work within the framework of on-going planning and decision-making efforts.

Notes

1. Sawicki and Peterman (2002).
2. East St. Louis Action Research Project http://www.eslarp.uiuc.edu
3. Neighborhood Knowledge Los Angeles, http://nkla.ucla.edu is affiliated with the University of California, Los Angeles (Also see Footnote 17, in Chap. 3).
4. Gale Cincotta and Shel Trapp were community organizers in Chicago. They co-founded the National People's Action (NPA), a coalition of community-based organizations and the National Training and Information Center (NTIC). These activists and the organizations they headed were instrumental securing passage of the Housing Mortgage Disclosure Act and the Community Reinvestment Act (Squires, 1992).
5. Richard Amanna, a graduate student in the Department of Urban Affairs and Planning at Hunter College, worked on many of these tasks under my direction in 2007–2008.

6. See http://rally-foundation.org
7. Open Space Accessible Information System (OASIS) http://www.oasisnyc.net/pages/about_OASIS.htm; Note that OASIS is now affiliated with a university, the City University of New York, Center for Urban Research.
8. The London Air Quality Network http://www.londonair.org.uk/london/asp/default.asp site allows users to understand the complex phenomena of air pollution monitoring, analysis and modeling over an extended time frame (1993–2007), with data now provided from 33 London boroughs. Users can display, graph, and download data about individual pollution parameters, for particular sites, and compare across sites. Additional information about London's Air Quality Strategy and target pollution reduction goals are also available for easy comparisons.
9. Living Independently in Los Angles (LILA) http://lila.ucla.edu/; LILA is a regional (county level) approach to addressing the needs of individuals living with disabilities in LA county. LILA includes a map room to assist local resources to create their own database based on their local "expert" knowledge to identify and map resources that support independent living.
10. http://gis.chicagopolice.org
11. Mashups are web-based applications that use data from multiple sources to create a new application to serve a particular purpose (see examples that follow).
12. Chicago Crime Map is a free browsable database of crimes in Chicago, with data gathered from the Chicago Police department and mapped using Google Maps Application Programming Interface, http://www.chicagocrime.org/
13. The Hiking Trail Database at http://www.trailheadfinder.com/
14. Hot Spotr, a community driven site that finds wifi hotspots at http://hotspotr.com/wifi
15. For example, 2007 Workshop on Volunteered Geographic Information http://www.ncgia.ucsb.edu/projects/vgi/
16. A review of web postings about the site indicates that the creators felt that the services offered by their site were more effectively provided by other entities including the City of Chicago itself.
17. Survey Monkey www.surveymonkey.com
18. The Institutional Review Board (IRB) is an independent compliance committee designed to protect participants in human research. It is mandated by the US Department of Health and Human Services (DHHS, Title 45, Part 46 of the Code of Federal Regulations). The IRB process is intended to protect the rights and welfare of individuals recruited to participate in research activities conducted under the auspices of Hunter College. At Hunter College, the IRB has the authority to approve, require modifications in, or disapprove all research activities that fall within its jurisdiction as specified by both the federal regulations and institutional policy. The materials and the survey protocols submitted by the research team were reviewed by a committee established for this purpose.
19. http://lila.ucla.edu
20. Jennifer Brisbane, personal communication.

Appendix: PPGIS Survey Instrument

PPGIS

Welcome: Public Participation Geographic Information Systems Survey Consent

Dear Colleague:

We invite you to participate in a short e-survey designed to understand the use and effectiveness of Geographic Information Systems that support public participation, an emerging field of study known as PPGIS (Public Participation Geographic Information Systems). This e-survey is part of research conducted by faculty and students at Hunter College of the City University of New York (Hunter).

Our research seeks to examine how PPGIS is applied in a variety of contexts to support and facilitate public participation. You have been contacted because we believe that your organization offers an application or service that provides the public with geospatial information in order to inform decision-making. Therefore, we invite you to participate in this research by responding to a brief e-survey that is expected to take about 10 minutes to complete.

Taking part is voluntary and your individual responses to all of the questions will remain confidential. Identifiable information is not required for participation.

While there are no direct benefits to you for participating in the survey, your considered responses to survey questions contribute to enhancing our understanding of the benefits and limits of GIS technologies that are designed to foster public participation.

We thank you in advance for your time.

Best regards,

Dr. Laxmi Ramasubramanian
Associate Professor, Urban Affairs and Planning
Hunter College of the City University of New York
695 Park Avenue
New York, NY 10065
212-772-5594
laxmi@hunter.cuny.edu

Richard Amanna
Graduate Research Assistant, Urban Affairs and Planning
Hunter College of the City University of New York
Richard.amanna@gmail.com

For information regarding the institutional approval process for this research, please contact:

Hunter College Institutional Review Board
Reference # HC-110713341
695 Park Avenue , Room E1426
New York, NY 10065
212-650-3053
http://www.hunter.cuny.edu/irb

Below, you are given two options:

By selecting "I'd like to participate; take me to the survey," you have consented to participate in this survey and will be taken directly to the questions.

PPGIS

By selecting "I'd like to see the full consent information before I begin," you will be taken to a page that details the methods and procedures employed in our research.

1. How would you like to proceed?

- ○ I'd like to particpate; take me to the survey
- ○ I'd like to see the full consent information before I begin

PPGIS

PPGIS Consent

Through a review of internet-based resources, your organization has been selected to participate in a research study of Public Participation Geographic Information Systems (PPGIS) by faculty and students in the Hunter College Department of Urban Affairs and Planning. PPGIS is a developing field that explores the relationship through which the everyday public can become involved with using, designing or creating geographic information systems. Your project has been selected based on its attempt to inform the public through the use of spatial data. It is anticipated that approximately 25-50 organizations will be surveyed during this phase of research.

Please complete a short survey that seeks to explore the nature of your organization and the organization's role in engaging the public in different aspects of decision-making. It should take about 10 minutes of your time. Taking this survey is completely voluntary and will be used solely for furthering academic knowledge about public participation and geographic information systems. If you decide not to participate at this survey, you may exit the survey at any time and your answers will not be recorded.

We are making every effort to ensure that no one knows what your responses were on the survey. Survey Monkey is a well-known company that collects data for online survey research. The study is not being run from a secure server like those used to handle credit card transactions, so there is a small possibility that responses could be viewed by unauthorized third parties, such as computer hackers. Please consult Survey Monkey's complete privacy policy by clicking on the "Privacy Statement" link at the bottom the company's home page (http://www.surveymonkey.com) Printed data collected by the researchers will be kept in a locked office cabinet, and only the undersigned researchers have access to your responses, via a password protected computer.

By clicking "next" and completing the survey, you are voluntarily submitting data to be used in academic research. No identifiable information is required from you, however, we do request that you select a description, from a list of choices, that best describes your organization. At the end of the survey, you have the option of providing contact information if you would like to be personally contacted for a further discussion of our research.

After the survey period is closed, on or about April 1, 2008, you may view the results on-line via a link that will be posted on this survey site.

For information regarding the institutional approval process for this research, please contact:

Hunter College Institutional Review Board
Reference # HC-110713341
695 Park Avenue , Room E1426
New York, NY 10065
212-650-3053
http://www.hunter.cuny.edu/irb

Please feel free to contact the undersigned researchers with any questions you may have or if you would like to be informed of more complete survey findings. Thank you for your participation.

Dr. Laxmi Ramasubramanian
Associate Professor, Urban Affairs and Planning
Hunter College of the City University of New York
212-772-5594
laxmi@hunter.cuny.edu

Richard Amanna
Graduate Research Assistant, Urban Affairs and Planning

PPGIS

Hunter College of the City University of New York
Richard.amanna@gmail.com

Click "Next" to take the survey!

PPGIS

Questions

1. Please select the statement that best describes your organization. (Select only one option)

- ○ Community-based data provider / data intermediary
- ○ Government-based data provider / data intermediary
- ○ Community Development Corporation (CDC)
- ○ Community-based Organization (CBO)
- ○ Non-Governmental Organization (NGO)
- ○ University-based research center
- ○ University-based community outreach or advocacy center
- ○ Policy research group
- ○ Other (please specify)

2. Which statement(s) best describe your goals in offering this GIS service or application? (Select all that apply)

- ☐ To offer information to members of the public
- ☐ To offer information to professional researchers
- ☐ To engage / encourage community involvement
- ☐ To advocate for community change
- ☐ Other (please specify)

3. Who is your primary target audience? (Select only one option)

- ○ Professionals
- ○ Community Organizers / Activists
- ○ Public
- ○ Other (please specify)

PPGIS

4. Who are the actual users of your GIS service or application? (Select all that apply)

- [ ] Same As Above
- [ ] Professionals
- [ ] Community Organizers / Activists
- [ ] Public
- [ ] Other (please specify)

5. Do you require users to create an account or login before they can use your service or application? (Select only one option)

- ○ Yes - Our service is free to use
- ○ Yes - Fee-based service
- ○ No
- ○ Other (please specify)

6. Which statement(s) best describe the sources of your data? (Select all that apply)

- [ ] Official Records (e.g. property data, environmental data, crime reports, etc.)
- [ ] Commercial Records (e.g. phone book listings, advertisements, etc.)
- [ ] Community Input (e.g. citizen opinions about favorite places in the community)
- [ ] Direct Observation (e.g. staff-recorded data gathered through field observations)
- [ ] Other (please specify)

7. Can users submit their own data?

- ○ Yes
- ○ No
- ○ Not applicable

PPGIS

8. Do you review user-submitted data for accuracy?

- ○ Yes
- ○ No
- ○ Not applicable

PPGIS

Comments

1. Please share any comments that you feel would help us better understand the nature, goals and/or purpose of your organization.

2. If you are open to having a longer conversation with our researchers about these topics, please provide contact information (phone or email) so that we may reach you. Your information will not be shared with any outside organization.

PPGIS

Thanks!

Thank you for taking the time to complete this survey. Your responses will help us better understand the emerging field of PPGIS. Remember, once the survey is closed on or about March 1, 2008, you can view our findings at a link that will be posted on this survey webpage. Also, for more detailed findings, please contact the researchers:

Dr. Laxmi Ramasubramanian
Associate Professor, Urban Affairs and Planning
Hunter College of the City University of New York
212-772-5594
laxmi@hunter.cuny.edu

Richard Amanna
Graduate Research Assistant, Urban Affairs and Planning
Hunter College of the City University of New York
Richard.amanna@gmail.com

Part II
Three Narratives

Chapter 5
Politics and Participation in Boston's South End

We may have come on many ships, but we are in the same boat now...

Mel King, long term resident of the South End

5.1 Introduction

This chapter tells a detailed story of a small coalition of housing advocates in Boston's South End. The coalition's participants are identified only by pseudonyms,[1] except for elected officials. The events of this story took place little over a decade ago – this is an important fact to keep in mind as you read further; this chapter is a history lesson in more ways than one. The events that are reported here, although they occurred in the 1990s, are in fact shaped by tumultuous events that occurred in an earlier era.

When people come together to solve societal (public) problems, the process is always political and it is always context dependent. This is what the practitioners stated, as discussed in the interview findings in the previous chapter. As you will see in the next section, the South End neighborhood of Boston has been at the center of many political struggles since the 1960s, and the organizers who are at the heart of this story were effective in integrating past and present, in order to create a coherent spatial story that was supported by the use of information technologies including GIS.

5.2 The City in History

The city of Boston, founded in 1630, accommodates close to 600,000 residents, while the Greater Boston area that includes cities like Cambridge, Somerville, and Brookline accounts for more than five million people, ranking it among the 10 most populous metropolitan regions in the country.

Boston is probably recognized universally as an important site in the American War of Independence. Like other American cities, the development of Boston in the

L. Ramasubramanian, *Geographic Information Science and Public Participation*, Advances in Geographic Information Science, DOI 10.1007/978-3-540-75401-5_5,

nineteenth and twentieth centuries can be seen as a direct outcome of technological development. Early on, innovative technologies and advances in transportation allowed planners and builders to create the Back Bay and the South End by filling in tidal mud-flats. Complex street networks and new streetcar lines defined the territorial expansion of the city. Decades later, innovative technological developments from the Massachusetts Institute of Technology and other universities would fuel the growth of hundreds of high-technology firms in the Boston metropolitan region.

Boston is a city of immigrants. The first wave of English settlers has been joined by the Irish, Italians, Blacks, Latinos, Asians, and Cape Verdeans among others. Presently approximately 45% of the city's population is comprised of people of color (Fig. 5.1).

Ambitious planning initiatives have characterized development in Boston. In the 1960s, the city overcame decades of economic stagnation by creating a pro-growth

Fig. 5.1 Map of the South End, Boston

partnership between the private and public sector. The Collins Administration, guided by its charismatic development administrator Ed Logue, created a powerful agency called the Boston Redevelopment Authority (BRA), which was responsible for city planning and implementing redevelopment and had the power of eminent domain.

The creation of the BRA was the cornerstone of "New Boston." This era was characterized by the initiation of several major redevelopment projects like the Prudential Center and the Government Center. Their completion changed Boston's cityscape permanently.

By 1965, the Boston Redevelopment Authority was addressing both physical and social development issues and advocated a city-wide strategy to "treat both causes and major symptoms of Boston's physical decline" (Kennedy, 1992, p. 185). One of the agency's goals was to reduce the concentration of low income people and non whites in the city's core while stabilizing the residential neighborhoods in Boston. The urban renewal program, supported by federal funding, destroyed many neighborhoods and communities such as the West End,[2] the South End and parts of Charlestown. These programs increased housing shortages, heightened racial tensions and caused many people to mistrust the motives of the BRA.

At the same time, city leaders and prominent bankers, who were concerned with preserving Boston's image as a viable center for economic development which was, by this time, marred by racial tensions, sought to expand home ownership opportunities for poor Blacks. Levine and Harmon (1992) drawing on interviews with long term residents argued that unfair banking practices created the physical and social segregation of Blacks and people of color.

> [The consortium] was complicit in establishing a carefully limited and well-defined inner-city district within which, and only within which, Blacks could obtain the attractive, federally insured housing loans. [This area] skirted the predominantly Irish and Italian working class neighborhoods and ... the suburbs where the bankers themselves lived (Levine and Harmon, 1992, p. 6).

5.3 The South End Resists Urban Renewal

Since the 1850s, when it was first planned, the South End was always a target of government-initiated planning efforts because of its prime location, near Back Bay with easy access to downtown Boston. Initially it was conceived as a high-income residential enclave, modeled after London, intended to counteract the exodus of wealthy Bostonians to the suburbs. However, as early as 1866, the South End had become a mixed-income neighborhood, accommodating among others an influx of poor residents relocated by urban renewal in the nearby Fort Hill area, as well as successive waves of immigrants.

In the 1950s and 1960s, city planners viewed the South End as a neighborhood in decline and ready to be "renewed", even though the residents (mostly Blacks, Hispanics, and immigrants), who considered it home would unlikely ever find suitable housing there again, given the prevailing social and economic conditions. The urban renewal projects proposed by city planners sought to remove "blighted areas" by targeting tenement houses and other housing options available to working class people. This removed about one fourth of the neighborhood's housing stock (in terms of dwelling units). Medoff and Sklar (1994, p. 20) noted,

> neighborhood tensions rose as the BRA's demolition work outstripped its promises of relocation and affordable housing. The tension wasn't over whether to renew the South End, but how and for whom?

Learning from its experiences with the West End Urban Renewal project (Gans, 1962), the BRA established a community participation process which it then co-opted – using the process to organize the community on behalf of its own bureaucratic interests. In addition, the BRA represented the interests of a small elite group of people who shared its own goals of transforming the South End into a "respectable middle-class neighborhood" (Mollenkopf, 1983, p. 175).

As a result, several community groups emerged to protest the BRA's policies and actions. Among them, an umbrella group called the Community Assembly for a United South End (CAUSE) began advocating for the rights of tenants and for South End residents with low and moderate incomes (King, 1981). The most famous example of resistance and protest against urban renewal policies in the South End was "Tent City," a spontaneous village that emerged on a 3-acre site that had been cleared of homes. The Tent City struggle began in April 1968 and was led by CAUSE organizer Mel King; he and other CAUSE members "occupied" a parcel of land earmarked for development by the BRA. After several struggles, the confrontations ended when a mixed income housing development rose on the site in 1987 (King, personal communication).

Medoff and Sklar (1994) comment that the Mayor Flynn adopted many of the neighborhood planning ideals advocated by his opponent (Mel King) in the highly charged election of 1983 including King's efforts to democratize neighborhood planning processes. The city began to advocate a Linked Development strategy in an attempt to join the prosperity of the downtown to the needs of poorer neighborhoods and people. This program required developers to commit funds to subsidize the creation of affordable housing. In 1986, the South End Neighborhood Initiative (SENHI) was launched. This endeavor was made possible with money from Linkage funds, BRA grants and loans, as well as real estate contributions, Community Development Block Grant loans from the City of Boston, in addition to funding from other private and public sources. Lawrence Kennedy, researching his 1992 book, *Planning the City upon a Hill* found that while residents in the South End continued to be wary of the BRA, the BRA's efforts to support and institutionalize community participation in planning processes in the late 1980s and early 1990s had helped to heal some very old, deep wounds.

5.4 The SEPHC Coalition

The South End Housing and Planning Coalition (SEHPC) was a grassroots action group that emerged almost as a direct response to the BRA's new planning strategies in the 1980s. The BRA created Planning and Zoning Advisory Committees to help formulate plans for individual districts. Instead of focusing on a master plan for the entire city, the BRA focused its attention on "comprehensive planning for each district and each districts' connection to the overall city" (Kennedy, 1992, p. 226). The agency established a community planning process in which city planners received the input of various neighborhood groups and then used this input to develop a "rational" plan for the district. The BRA also instituted a public review process, in which developers presented their design schemes and alternatives to meet public scrutiny and approval.

SEHPC members became active in monitoring community planning processes in the South End. For instance, they monitored the compliance of developers with previously established commitments such as the city's commitment to ensure the creation of low and moderate income housing within each new development. It is within this framework that SEHPC became involved with the planning of the proposed development called the South End Technology Square (SETSA), a multi-year, multi-phase construction project initiated by a consortium of private developers, including Boston University.

5.5 The SEHPC Challenge to SETSA

SETSA was to a bio-medical facilities building. Its development was promoted by the Flynn administration in the early 1990s under the premise that investments in biotechnology and biomedical research would help to generate new jobs for Boston's unskilled and semi-skilled populations clustered in the neighborhoods of South End/Roxbury.

While it was readily apparent that successful economic development initiatives like SETSA would benefit the city's coffers and enhance Boston's reputation, SEHPC members were less sure about the new development's potential to generate jobs for South End residents. In order to educate themselves, they hired a neighborhood planner and asked him to investigate the issue on their behalf. They challenged the planner to ask hard questions, including questions about the number of low- and semi-skilled jobs that would become available; the potential for small business enterprises to establish linkages to larger biomedical enterprises; and whether the new developments would trigger a process of gentrification. After sifting through a considerable volume of data, the planner concluded that:

> the expansion of the biomedical industry in Boston will improve the city's economy but its corresponding job growth *will not benefit* the majority of Boston residents who are in need of jobs because the educational level of [these] residents will not match the educational requirements of the biomedical industry (SEPHC publication, 1991, Executive Summary).

Armed with this new information, SEHPC successfully argued that the new jobs (production and manufacturing jobs) that would have traditionally used low skilled workers were more likely to shift to the Third World or other areas of the country like North Carolina that offered cheaper land and labor and other economic incentives, thereby eliminating potential community-wide benefits that were likely to result because of new job creation.

In this instance, socio-demographic data from the US census coupled with additional data from economic censuses organized by census tracts was used creatively by the SEHPC coalition, to negotiate for community benefits in the form of a new community health center to serve the needs of the South End community. As the leader of that coalition commented to me in an interview I conducted with him in 1996.

> They [developers], needed a zoning variance, so we had a hook to get some community benefits before we gave our support. . . .we knew [from our research] that we would not get any jobs from the [biomedical] project, so we did something, we asked them [the developers] to use their influence to help locate and finance a new state-of-the-art community health center [Note: This center has since been built].

SEHPC members moved away from their original arguments. Initially they began negotiating for a percentage of the new jobs to be earmarked for qualified local residents, a conventional approach in negotiating community benefits.[3] Once they learned through their own research and analysis that the new jobs would not immediately and directly benefit the most vulnerable residents in their neighborhood, they began to advocate for the creation of a new community-health center that would be financed and built by the developers to serve the needs of all residents in the neighborhood. By identifying the jobs-skills mismatch early on, the coalition was able to establish new ground rules for the negotiation of the community benefits agreements.

5.6 How SEHPC Works

In Chap. 1, I observed that any attempt to integrate public participation in planning has to confront four major dilemmas – in the framing of planning problems, in determining the locus of planning authority, in determining the public interest, and in managing the participation within formal planning processes. I argue that the South End Housing and Planning Coalition (SEPHC) was able to adroitly navigate these dilemmas because of the political savvy exhibited by their members, in particular, its leaders, the coalition's commitment to the principles of community organizing, and because they had ready access to highly customized and customizable geo-spatial data.

According to the SEHPC Executive Director, Mr. Harrison, the coalition's goal, advocacy for institutional change, requires the group to find a way to shape public policy on behalf of low income people and people of color in the South End. However, he was quick to point out that "[in terms of public policy debates], *we're kind of judicious, we don't get involved in every tiny little thing*." SEHPC does not enter the public arena of confrontation easily or quickly. The process begins by building a coalition of grass-roots support and their negotiations with different interest groups are backed by organizing and knowledge.

According to the Executive Director, technologies are integral to do the kind of research that is needed to make compelling arguments. He says:

> I am convinced that organizing in the future should be a marriage of technology and old-fashioned community organizing. I don't think technology, without the organizing, is going to work. And I think, organizing, without the technology, in the '90s and in the new century, is not gonna work.

Mr. Harrison recognizes that the availability and use of technologies is only one aspect of the entire research process. He emphasizes "analysis" which, in his view, is somewhat synonymous with research. "Lots of people are caught up on the means, the means are not the answer to everything. You still need a lot of analysis".

SEHPC is an inclusive organization that addresses the concerns of South End residents, neighborhood groups, as well as the interests of various for-profit and not-for-profit businesses that are active in the community. While it has a historical reputation of being a spokesperson for the under dog, in recent years several influential players including the leadership of the City of Boston, have come to view the SEHPC coalition as a strategic player that can be invited to work in partnership with the City and the private sector in order to accomplish many community development goals. SEHPC's reputation and position of influence have evolved over time, partly because SEHPC has consistently won its public battles with the City and other powerful interest groups. The fact that elected officials of the Massachusetts State Legislature representing the South End, as well as some members of the Boston City Council and many other community leaders participate in coalition meetings adds to the cache of the coalition.

Some of SEHPC's power and name recognition are directly associated with its energetic Executive Director, Mr. Harrison. He has a long history of community activism and has participated in the development of the South End for several years. One of the residents interviewed for this research noted, "we've known Mr. Harrison [the Executive Director] for quite a long time . . . his history of marching with Martin Luther King . . . his experiences [as an organizer]. . ."

This coalition's coordinating committee, a group of about a dozen individuals, regularly attends planning and strategy meetings. The small number of people at these meetings is conducive to facilitating participation. This group includes executive directors or senior staff from neighborhood organizations, representatives of homeowners and renters associations, and representatives of other single issue (e.g., urban gardens) groups. Many of the participants are also long term South End residents.

Coalition members know each other fairly well since they have worked together over a period of time,[4] addressing a variety of issues. Ms. Cochran, a long time South End resident explained, ". . . we've [referring to the members of the group] known each other for a long time. . ." as she reflected on the frank discussions and exchanges that usually take place at coalition meetings. Mr. Harrison added, "some of the members have worked together for over ten years, so . . . I, as a human being, can absorb by looking at Jim, [a coalition member] very quickly as he's saying something ... knowing him and have interacted [with him] for ten years ... there's a whole lot of synthesis that goes on ... it's not intuition".

Most members of the SEHPC coalition observed that they were comfortable with the use of new information technologies. They emphasized the point that they were familiar with data analysis concepts and were thinking spatially even before technologies like GIS were introduced into the organization. However, they particularly appreciated the maps generated through spatial analyses, because the maps could be customized to suit organizational needs.

SEHPC has been using data and information extensively over 10 years or so it has been in existence. Mr. Harrison, Executive Director, proudly pointed out that, "the coalition has not lost a single issue in ten years before the BRA in terms of public policy..." He attributed much of this success to the judicious use of data and information. He was almost gleeful when he talked about "customized data bases and maps with graphics that were better than those produced by the City," and analyses that went well beyond counting numbers of people per census tract.

This organizing/advocacy role is usually played out in a politically charged environment and tends to be reactive. SEHPC said that they used data and information to challenge authority, but it must be noted that they aspired to move the conversation to a more pragmatic discussion of community benefits. The group was not anti-development, yet they were focused on supporting development initiatives that would improve the quality of life for the neighborhood residents.

Individuals or groups that engaged in confrontations with city agencies were unlikely to wholeheartedly trust the data offered by that agency. SEHPC often conducted its own research or double checked data and information provided by outside agencies. Their work affirms Davidoff's earlier comments about the benefits of advocacy planning, that is to reveal and challenge sloppy analyses on the part of professional planners working for government.

Data was often used in strategy sessions to assist SEPHC members in their short term and long term planning. Maps and powerful graphics were sometimes used to inform, educate, and attract the attention of residents and outsiders towards the work of the group. A South End resident, Ms. Cochran, observed:

> [Maps] put into graphic form some of the stuff we know, or don't always know, about what's going on around us. The older maps are nice because sometimes they show the configuration of the housing before they took it all away. Its a kind of history... Then there are those maps that go way back and show changes from various different times... (Ms. Cochran's Interview, 1996).

Mr. Harrison and Mr. Sivakumaran from SEHPC are ardent proponents of data visualization to support the communication and discussion of planning issues.

> Most people in other neighborhoods and community organizations don't realize the value of graphics. Mr. Sivakumaran did a chart ... I remember, in the 1980s ... it showed an affordability gap in this neighborhood between income and cost of housing. And it showed [trends] over several decades. Well, see. I retain [the map] in my mind's eye because of the graphics... (Mr. Harrison's Interview, 1996).

> Maps are a representation of reality. I customize maps, I include [qualitative] information, pictures, and integrate data and statistics with issues... like crime, like housing (Mr. Sivakumaran's Interview, 1996).

It is important to note that the actual research about the nature of the biotechnology industry, including an investigation of the economic impacts of bio-tech related investments in other United States cities was done by a professional planner (an expert) with access to extensive databases and research reports and thorough familiarity with demographic analysis. What is unique is that the SEHPC coalition members (everyday folks) were actively involved in framing the questions that guided the planner's research, and were actively involved in analyzing the data and results that emerged.

From this experience, I concluded that it is very important for community residents to be engaged in framing the questions and for them to actively immerse themselves in the discussion of the analyses and the results. While participating in data collection (community mapping) is a useful activity, the success of a participatory planning/research endeavor relies on having the access, background knowledge, and the power to ask questions – in other words, being able to interrogate the data critically.

Notes

1. At the time I conducted the interviews (1996–1997), I did not seek written consent to use the names of the organization and the participants. Hence all names used in this narrative, except the names of elected officials are pseudonyms. Some of the original participants are since deceased.
2. Gans (1962).
3. Annie E. Casey Foundation, Baltimore, Maryland, 2007. Community Benefits Agreements: The power, practice and promise of a responsible redevelopment tool.
4. At the time the data was collected, 1995–1997, coalition meetings were usually held regularly, for example, monthly. However, the number of meetings increased when there is an urgent problem or issue that needed attention.

Chapter 6
Planning to Preserve Community Character in Oak Park, Illinois

The deepest meaning of any place is its sense of connection to human life and indeed to the whole web of living things.
Kevin Lynch, Managing the Sense of a Region

6.1 Introduction

This chapter describes my own personal experiences working on a project funded by the Village of Oak Park, Illinois. The project itself was a year-long collaboration between the Village of Oak Park, Illinois and the University of Illinois at Chicago (UIC) in which a fairly sizable group of faculty, graduate students, village staff, citizen activists and volunteers came together to develop neighborhood character plans for two commercial business districts in the village. One of the unique contributions I made to this project was to explore how computer-mediated visualization and communication tools could be used to complement and facilitate conventional community organizing and traditional participatory planning.

The collaborative project henceforth referred to as the Oak Park project highlights both advantages and limitations of doing participatory planning through university-community partnership projects. The analysis of the state of the practice in Chap. 4 revealed that many PPGIS projects are housed at, or otherwise affiliated with universities or university-based research centers. Prominent examples of such on-going partnerships include the East St. Louis Action Research Project (ESLARP) affiliated with the University of Illinois at Urbana-Champaign and Neighborhood Knowledge Los Angeles (NKLA) and Neighborhood Knowledge California (NKCA), and Living Independently in Los Angeles (LILA) all affiliated with the University of California at Los Angeles. As the Oak Park case study will show, PPGIS adoption and use appears to simultaneously benefit from and be constrained by the academic enterprise.

The Oak Park case study scales up our discussions about participatory planning using digital tools. While the South End case study (Chap. 5) addressed planning issues in one neighborhood, this case study reports on efforts to engage the entire

L. Ramasubramanian, *Geographic Information Science and Public Participation*, Advances in Geographic Information Science, DOI 10.1007/978-3-540-75401-5_6,

village in proactive conversations about development and change. Although the village's population is relatively small (approximately 50,000 people), the idea that participatory planning can simultaneously address neighborhood level concerns and village-wide concerns is worthy of critical investigation. The Oak Park case also highlights many of the challenges associated with institutionalizing participatory planning and embedding these activities within day-to-day planning practice.

After describing the project activities (Sects. 6.2, 6.3, 6.4, 6.5 and 6.6), I reflect on the experience to assess how the Oak Park Project transformed planning processes in the Village. I also draw conclusions about the transformation of both public participation methods and GIS techniques in the 6-year gap between 1996 and 2002, i.e., when the technology-supported participation work undertaken by the South End Housing and Planning coalition described in Chap. 4 ended and when the Oak Park project began. With this background, let us move forward to learn more about the Village of Oak Park in Illinois.

6.2 Oak Park, Illinois: Background and History

The Village of Oak Park was incorporated as a municipality in 1902. Because of its proximity to Chicago and its rich transportation connections, Oak Park was, and is a bedroom community, serving both the City of Chicago and the outer suburbs. Frank Lloyd Wright (FLW), one of America's most famous architects lived and worked in Oak Park, and the Village contains the largest collection of FLW designed homes. Driving or walking through Oak Park, visitors are often struck by the preponderance of Prairie Style homes, the design that Wright made popular. In addition, the Village contains several landmarks, including the Unity Temple, Wright's first public building, and Wright's own residence and studio.

Oak Park, is very self conscious about its unique identity, perhaps because of its need to compete with its bigger and louder neighbor (the City of Chicago). The Village's website states:

> Oak Park is a thriving community of about 52,000 people located immediately west of the City of Chicago and known for its architectural heritage and diverse population. Within its 4.5 square miles live one of the region's most diverse mixes of cultures, races, ethnicities, professions, lifestyles, religions, ages and incomes[1] (Fig. 6.1).

According to the 2000 Census, Oak Park's profile looks like this. Of persons who selected to identify only one race about (95% of the population), 69% are White, 22% are Black/African American and about 4% are Asian. Hispanics or Latinos (of any race) make up about 4% of the population. Considering that the Blacks/African Americans make up about 12.4% of the national population, Oak Park is doing very well in terms of racial diversity. However, in terms of economic diversity, professional diversity, and age diversity, the claim is slightly overstated. The median household income for Oak Parkers in 1999 dollars was $59,183, significantly higher than the US as whole ($41,994). Oak Park residents are also better educated, over 62% have at least a bachelor's degree compared with 24.4% nationwide. In 2000,

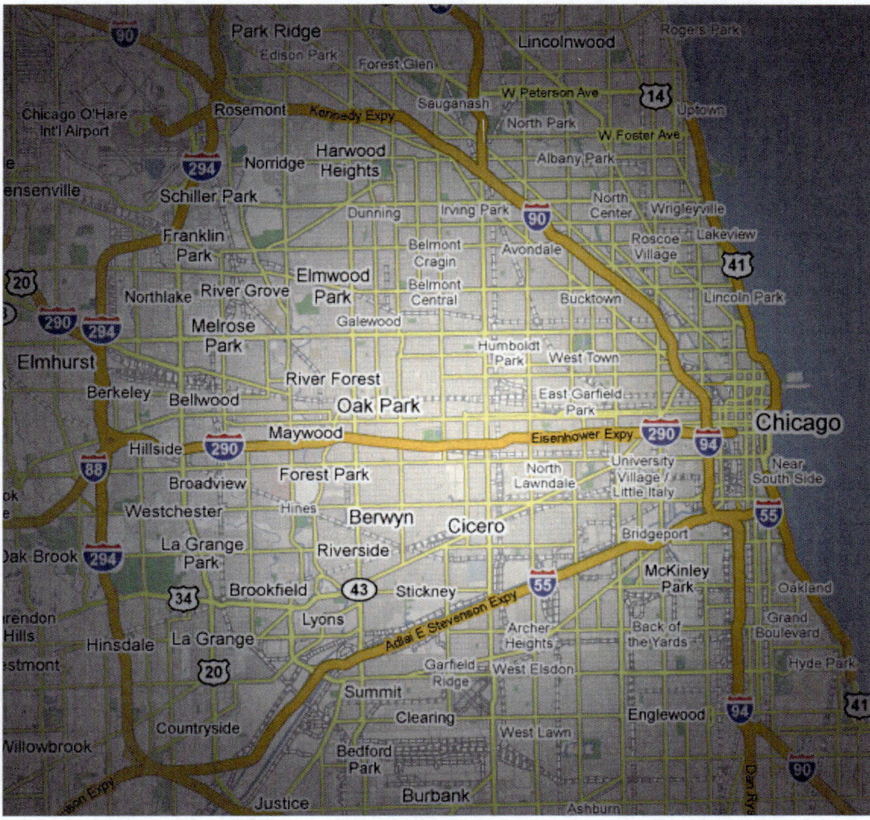

Fig. 6.1 Map of Oak Park, Illinois

the median age of Oak Park residents was 36 years, with only 9.5% of the residents being 65 years or older, lower than the national figure of 12.4%. The housing stock in the village, approximately 24,000 housing units in total, is more or less evenly distributed between owner-occupied homes (56% of occupied housing units) and rental units (43.7% of occupied housing units). In 2002, the average single-family home in Oak Park cost $231,300 (US Census 2000). The 2005–2007 American Community Survey 3-Year estimates reveals that socio-demographic profile has not changed dramatically.

Oak Park is governed by a Board of Trustees led by the Village President (elected representatives) who set the policy agenda. The trustees appoint a Village Manager (typically someone with planning and public administration experience) to run day-to-day Village affairs. The Village Board has consistently used policy instruments to create a racially and economically diverse community. For instance, Oak Park is one of the earliest communities in Illinois that passed a fair housing ordinance in 1968 and has worked carefully and proactively to sustain residential integration despite numerous difficulties (Squires, Bennett, McCourt, & Nyden, 1989). The nationally recognized diversity statement adopted by the Village of Oak Park[2] underscores its commitment to continue to support its fair housing philosophy. This philosophy requires that housing opportunities are offered equally to all persons, regardless of race, economic status, gender, age, ethnicity, sexual orientation, disability, religion, political affiliation, or any of the other distinguishing characteristics that tend to divide people in society.

6.3 Planning in Oak Park

In its *1990 Comprehensive Plan*,[3] the Village set out general goals related to housing, transportation, and parking; public facilities; economic development; and public participation. This *Plan* is a long-range policy guide for the future physical and social development of the Village. The *1990 Comprehensive Plan* articulates the community's commitment to human values; a sense that the Village exists for its citizens; that the physical manifestation – housing, parks, schools, businesses, etc. – are there to meet the needs of its constituents.

The 1990 Plan developed goals in five key areas – housing, transportation and parking, public facilities and services, economic development, and citizen participation. Most of these elements are ingredients of most comprehensive plans. Yet, Oak Park's plan distinguishes itself by explicitly articulating its commitment to the:

- provision of affordable housing;
- preservation of the residential character of its neighborhoods;
- support for a broad range of convenient retail and service facilities; and
- high level of citizen involvement in village affairs.

The Oak Park diversity statement, adopted by the village president and the board of trustees in May 2005 is a further affirmation of that original commitment to social integration as well as an explicit commitment to support active citizen engagement in planning.

> Oak Park's proud traditions of citizen involvement and accessible local government challenge us to show others how such a community can embrace change while still respecting and preserving the best of the past. . . Our goal is for people of widely differing backgrounds to do more than live next to one another. Through interaction, we believe we can reconcile the apparent paradox of appreciating and even celebrating our differences while at the same time developing consensus on a shared vision for the future. Oak Park recognizes that a free, open and inclusive community is achieved through full and broad participation of all its citizenry. We believe the best decisions are made when everyone is represented in decision-making and power is shared collectively.

To summarize, it is reasonable to observe that Oak Park is a community that values citizen engagement in planning activities.

6.4 Coping with Growth and Change

At the time the Oak Park study was being discussed in early 2002, both Village elected officials and residents were becoming concerned with the growth of the Chicago metropolitan region[4] and the consequent impacts on Oak Park. The village was facing intense development pressures pitting proponents of economic development (who argued for increasing residential density as well as the diversity of housing options) against those who advocated a more incremental growth strategy. A community group, Responsible Economic Development Citizens of Oak Park (REDCOOP) was very vocal in their criticism of the village's approach to property development, particularly village-owned properties. Members of this group indicated that they were particularly concerned about the lack of transparency. Nile Wendorf, president of REDCOOP declared,

> When the village decides to develop property now, many of the key decisions are made behind closed doors. The request for proposals is drafted without public input, it's sent to developers without public input, and the chosen proposal is selected without public input.[5]

He scolded the elected officials and village staff, "you should get your citizens involved before you go into the RFP process...",[6] although REDCOOP offered no specifics about how this should be done. Initial opposition to the village's approach to development was focused on Whiteco Residential LLC's proposal[7] to develop a publicly owned parcel, an under-used parking lot at Harlem and Ontario streets with a view to creating a residential complex with a high end grocery store, health club, and abundant parking. Most Oak Park residents were concerned about the scale and density of the development.

At around this time, the Village trustees invited the University of Illinois-Chicago to develop a proposal to study two of the village's 12 business districts. These two to four block stretches of small stores and businesses were spread out through the village. In many cases, they were utilitarian but non-descript, and in some instances, were quite run down (Fig. 6.2).

In keeping with the Village's politically and socially progressive roots, and its commitment to open and inclusive planning, the Village trustees and the Village manager invited the university to create a process that was innovative and inclusive in order to facilitate a broad conversation about neighborhood character in the context of revitalizing the Village's retail business districts.

The university was charged with delivering a product, namely recommendations to revitalize two business districts, the Oak Park-Eisenhower Avenue retail business district and the Harrison Street retail business district. For the village trustees, the purpose behind this particular selection of business districts was to help bolster the village's case against possible expansion of the Eisenhower Expressway (the Ike), an Illinois Department of Transportation (IDOT) proposal that was in early stages of planning around the time the UIC contract was awarded.

In addition to specific recommendations for the two business districts, the university was also charged with providing a set of tools and frameworks to facilitate participatory planning. Charles Hoch, the UIC planning program director served as

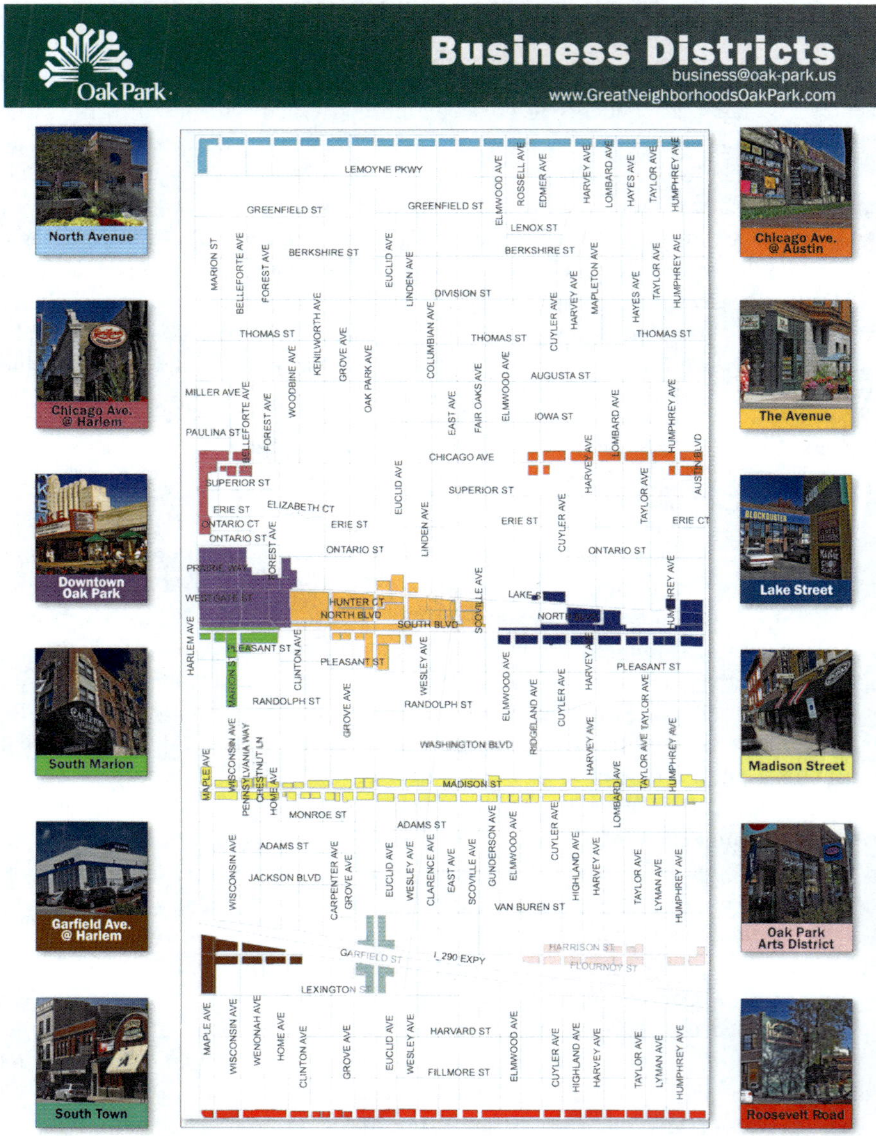

Fig. 6.2 Map of Village showing retail business districts

the project lead. He touted the benefits of using digital tools at an early stage of the project. "People who can't come to a meeting can still log on (to the Internet from a personal computer) and give their opinions",[8] he said.

For the trustees and the village staff, the alliance with the university became an elegant way to seek advice on how to guide development using participatory processes from a group of local experts, while simultaneously defusing some of the community criticism directed against them at the time. The village's concern about IDOT's expressway expansion plans and their desire to counter any negative consequences associated with this expansion may have played a part in moving the UIC-Oak Park partnership project forward. In any event, the project was formally launched with a community meeting in September 2002.

The tensions that were bubbling under the surface came to the fore almost immediately. Attendance at the first community meeting about the project was high. There were many curiosity seekers who wondered what UIC's role was going to be. Residents and business owners who lived and worked in the target planning areas showed up and were quite anxious – there was a general perception that the Village Board was pro-development and many people had come prepared to voice their strong opposition. The UIC team used that first meeting as a listening session, establishing only the minimum of ground rules and opening up discussions for community input. Comments from stakeholders logged through the project website and in community meetings in those early days articulated some of these tensions.

> We hope that the charm and diversity of Oak Park will not be replaced by a slicked-up corporate mall look.
>
> I attended the meeting last night. . . I had the feeling that the tenor of the group was anti-high density. Just to let you know that there are residents of Oak Park who have other opinions. [Areas] close to the "El" [public transit] make good candidates as areas for high density housing.
>
> We should not recreate downtown Oak Park on Harrison. Businesses on a smaller scale – restaurants, used book stores, music stores, along with practical businesses like hardware or grocery stores (as long as not a huge chain) – seem appropriate.

6.5 An Overview of "Planning Together 2002–2003"

The goal of the Oak Park project (the subject of this case study) was to develop character plans for two business districts in two distinct neighborhoods within the village. While the process was initially initiated by a range of citizens, who had argued for the need for such proactive planning to spur economic development in these two neighborhoods, it was the elected representatives (the Village Trustees) who eventually moved the participatory action research agenda forward by inviting the local university (University of Illinois-Chicago) to design and implement the planning process, rather than hiring a planning consultant to deliver a particular product.

Both retail business districts selected for analysis were near public transportation (specifically adjoining transit stations). In the case of the Oak Park Avenue Retail Business District (henceforth Oak Park Ave.), the business district is divided by the Eisenhower Expressway. A commuter train from downtown Chicago (the "El") runs on the median of the expressway and is accessible from the bridge that runs over the expressway and that divides the retail area. The retail business district included a range of stores including a grocery store, a laundromat, a coffee shop, an arts and crafts store, and an Ale House. While generally popular in the community, the

Fig. 6.3 View of Oak Park avenue retail district

Fig. 6.4 View of Harrison avenue retail district

Ale House had attracted a fair share of criticism related to noise and public health concerns. Oak Park Avenue is accessible by both public and private transportation (Figs. 6.3 and 6.4).

The Harrison Street district is an interesting mix of single-family homes, multi-family apartment buildings, and commercial storefronts, many of which are occupied by various kinds of arts-related businesses. The street itself is a pleasant tree lined street with sidewalks, accessible by both public and private transportation. Harrison Street is a gateway to Oak Park because it has an off-ramp from the Eisenhower Expressway. The El train previously described also stops within a 5-min walking distance of Harrison Avenue.

The project activities and timeline reveals the complexity of the public participation activities undertaken (Fig. 6.5). They include:

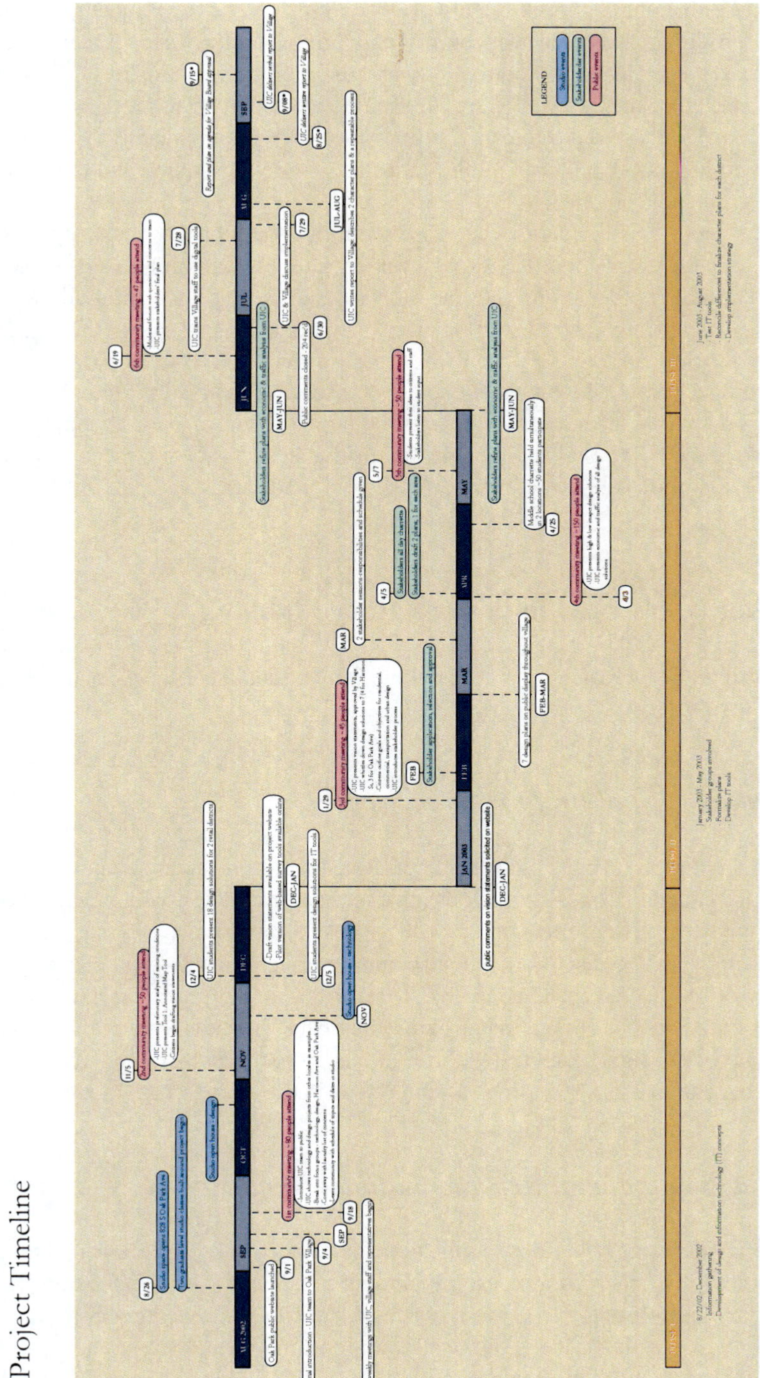

Fig. 6.5 Oak Park project timeline

- Public Meetings held approximately every two months of the project's duration, held at the Village Hall. These meetings including the initial kick-off meeting that introduced the project to the community and the final presentation of findings that served as a culminating event were informational. Each public meeting typically lasted 2 h. Synthesis and notes from each meeting were available to the public via the project's website. The public meetings were video-taped and ran on Oak Park's local cable channel.
- Meetings with stakeholders, who were identified through an open application process managed by the UIC team. Over 75 individuals volunteered to participate. The stakeholder group met face-to-face and virtually (over email) several times between March and May 2003. They also played a lead role in the charrette described below.
- A planning charrette at which stakeholders presented the first draft plans they had developed with assistance from experts. A larger group of participants worked intensively with the stakeholders to confirm the accuracy of the analysis of existing conditions and to refine and shape design ideas.

6.6 Digital Tools Used in Participatory Planning

Several different types of interactive digital applications were developed over the course of the project.

6.6.1 Online Visual Preference Surveys

Citizens benefit from conversations about existing conditions of any planning area as they plan for change. Typically, citizens have these conversations at a meeting without the support of visual information. The visual preferences survey can facilitate a more comprehensive discussion about planning issues such as character, density, and safety concerns. It can be adapted for use in group settings or it can function like a traditional on-line survey that provides feedback directly to the planning staff. The two surveys designed and tested through this project provided citizens a simple way to convey their ideas regarding design issues in the study areas (Fig. 6.6).

6.6.2 Navigational and Representational Applications

Navigational and representational applications allow citizens to take a virtual tour of these business districts. From a design/planning perspective, this application facilitates virtual walk-throughs, akin to a conventional site visit. Using a conventional point and click user interface, users walking along the virtual street, can pause at specific vantage points to get a sense of what lies ahead and look back to where they have just come from. They can get a 360-degree panoramic view of the area as well.

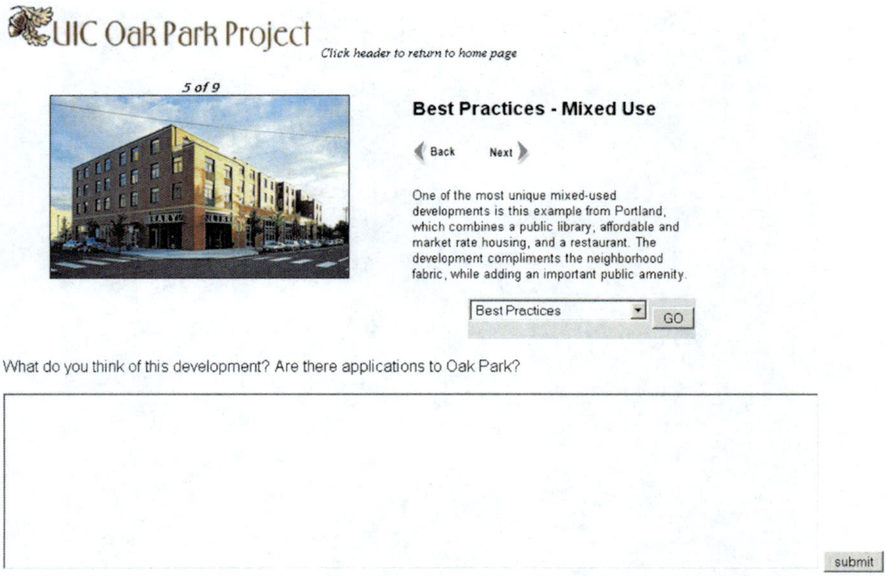

Fig. 6.6 Visual preference survey

This navigation tool facilitates discussion among stakeholders about existing conditions and can be further adapted to incorporate proposed changes and modifications (Fig. 6.7).

6.6.3 Annotated Maps

Aerial views of each business district served as a base map for the project. UIC students subsequently added a walk-through of the street embedded with visual cues (images and 360-degree panoramas) and text notes (Shiffer, 1995). This application was designed to be available on a stand alone computer (such as in a library or a community kiosk), where different comments could be saved and made visible (if desired) to the other users. During the project, this application was available for use by citizens in the studio. The "annotated" map could be integrated into a conventional power point presentation. This application was designed to be easily integrated into the day-to-day planning work of the Village. Village staff were trained in the use of this application and felt that it could be used for making presentations to the Board, at meetings with developers, and in meetings with citizen groups.

6.6.4 Online Planning Tools

Online planning tools allow users to communicate information to decision makers and to other citizens over the Internet without the use of any specialized software.

Fig. 6.7 View of Oak Park avenue business corridor straddling the freeway

These applications, when used in the context of a planning project allow decision makers to gather data about specific neighborhood level planning issues as and when feedback is needed. Used carefully, these on-line planning tools can complement and enrich participatory processes.

The two online tools customized for use in this project are modeled after applications developed earlier at the Urban Data Visualization Lab[9] at UIC. These applications allow users to draw on maps and submit comments accompanying their drawings to the planning team. This application provides users with a means of singling out specific intersections, blocks, or areas that require further scrutiny in the planning project and communicating the details of their concerns to the planning team. The Sketch Tool is available to users over the Web. In each instance, users are presented with a base map of the study area. They can select an area of the map by choosing to draw with a line, a point, or a rectangle shape. Once they have selected an area on the map, they are then asked to type comments which correspond to the area that they have selected. They submit this information which is saved in a database. They can also see "other views", which shows them areas that other users have selected and the comments that were submitted. These applications were developed during the course of the planning project and were presented to the public in April 2003. By changing the base map, the Sketch tool can be easily used in other geographical locations. The application could also have creative applications for other projects, where the Village solicits citizens to submit ideas with a place-specific aspect.

6.7 Benefits and Constraints

Innovations in electronic communication and visualization offer great promise to enhance citizen participation. However, they cannot and should not be seen as a complete substitute for face to face meetings or other forms of direct citizen involvement. When used creatively, these technologies can improve the quality and the efficiency of public discussions and debates and help build community consensus around specific planning issues.

In a pragmatic sense, the use of these interactive applications can increase participation among those citizens who are unable to attend face to face meetings. However, the tools collectively offer additional advantages. These applications protect the privacy of respondents and therefore allow citizens to share "unpopular" or "minority" opinions without the fear of personal attacks or criticism. For instance, the survey of existing conditions contained an image of a neighborhood ale house, which was also perceived as an unwanted land use by some residents, who lived in the area. The web-based survey suggested that a majority of respondents actually appreciated the ale house and frequented it regularly! The outcomes of the survey do not imply that there are no problems associated with the ale house. However, the issues that were brought up in the survey concerned noise pollution, sanitation (in the alley behind the ale house), and lack of parking – all planning issues that

Fig. 6.8 Land use map with image of ale house

can be managed without having to close down the facility. The level of detail provides additional information to planners and decision-makers, who are considering approvals of similar land uses in the same area (Fig. 6.8).

Interactive applications make it possible for users to become proactive rather than just reactive in thinking about the future of their community. The applications developed for these two business districts can be readily adapted for use in other business districts. The interactive tools are likely to place an additional burden on limited Village staff resources since the presence of interactive technologies are likely to raise expectations among citizens who anticipate personalized and immediate responses to questions and complaints. Citizens and users need to be educated about the value and benefits associated with these technologies. Although the Oak Park community has a high level of access to technology, some people are likely to feel overwhelmed and intimidated. The adoption and use of digital technologies must be accompanied with educational support and training programs for citizens to use the new tools.

6.8 Changes in Oak Park's Planning Approaches

The project concluded in August 2003. The village board trustees voted to receive the character plans and directed the Plan Commission to review the plan's recommendations. At a hearing about the plans, the chair of the Plan Commission reported that, "among those who participated, there was "absolute consensus" that it (the year-long effort) was a wonderful process and that it worked well. Many people came together to try to attain consensus about some difficult issues; it was a positive process" (Village of Oak Park, 2004).

The project planning team developed a series of interactive digital applications to meet the needs of the project and to demonstrate innovative use of technologies – these include web-based visual preference surveys; online sketch planning tools; planning portals; and a project website. Collectively the digital applications assisted in envisioning the immediate and long term future for these two neighborhoods, discussing the pros and cons of particular planning changes. For example, a high impact scenario visually and quantitatively showed how new development could be scaled up to generate new tax revenues that could benefit the village as a whole while highlighting the quality of life issues (traffic, displacement etc.) that would be compromised in the immediate vicinity of the development in pursuit of these goals.

From the university's perspective, the unique contributions of this project were:

- the use of different computer-mediated communication and visualization techniques to facilitate specific urban design and planning activities at the micro-neighborhood scale;
- the emphasis and attention placed on participatory process itself, in which faculty and staff invested time and effort designing and implementing a genuinely participatory process that was infused at different stages with innovative electronic technologies;
- the development of planning capacity (essential to sustainable development), among the citizenry and the planning staff at the Village;
- the creation of an stakeholder group representing different community interests who continue to work collaboratively to champion and shape the development of two retail business districts; and,
- the inclusion of voices not typically heard in participatory planning processes (e.g., youth, elderly, renters, and self employed people).

Digital technologies reinvigorated an already engaged community by creating pro-active (rather than reactive) participatory planning processes. For example, the project team was quickly able to show different types of project alternatives – the so-called low impact scenarios and the high impact scenarios. Eventually, stakeholder groups developed their own scenario that included elements from both options presented by the university team.

In addition, some of the techniques that were used to revitalize the participatory process were made possible because of the team's decision to use computer-mediated communication and visualization technologies. For instance, the use of a safe/neutral electronic space (the project website) allowing citizens to post anonymous comments made it possible for citizens to share their real thoughts, for example about high density. The team quickly learned that not all Oak Parkers were anti-high density, contrary to anecdotal evidence, but instead learned all citizens were very concerned about having an open planning process. Likewise, the team was able to post documents and plans (work in progress) on the project website to show how ideas about the planning and design of the retail business districts evolved over time. In some instances, the technology was invisible to the end user. For example, the team used GIS as a back-end data processing and analysis tool.

Traditional GIS-generated artifacts such as high-quality maps, and analyses of cadastral data were integrated into easy-to-use document formats such as PDF files and PowerPoint. Overall, the project demonstrates how computer-mediated communication and visualization technologies can be used to enhance traditional participatory planning. It requires extra time and a lot of patience, but an engaged citizenry aids the planning process.

6.9 Conclusions

In February 2006, about 2 years after the Planning Together project concluded, the Village of Oak Park developed Guidelines and Procedures for Participatory Planning that govern the development or re-development of any Village-owned land (Village of Oak Park, Board of Trustees Policy, 2006). The guidelines state:

> the purpose of creating the public participatory planning guidelines is to ensure that each village-owned property being considered for development/ redevelopment is reviewed in a consistent and open manner...

The guidelines emphasize open communication and the need to raise awareness about planning issues in the Village across a wide swath of the public and the need to provide multiple opportunities for review and comment. The UIC-Oak Park project showed elected officials and planning staff that most citizens understood the need to make trade-offs and were able to balance their interests and commitments to maintaining community character with the needs of growth and economic development.

Even more rewarding is the realization that the Village's current plan for the redevelopment of one of the districts (the Harrison Street Arts District) in Oak Park developed by the Lakota Group,[10] a planning consulting firm, incorporates many of the key design and planning recommendations[11] made by the Oak Park project. That the UIC team of faculty and graduate students was able to help visualize the design for the district that eventually incorporated into an implementation plan is additional confirmation that the Oak Park project process was a credible one.

The phrase "neighborhood character" is often a code word that captures individual preferences about place. Neighborhood character incorporates many elements such as density, height of structures, land use, availability of public transportation, and traffic. Neighborhood character can also be influenced by socio-economic and racial diversity, and the presence or absence of activities and services that can meet individual needs. Planning elements can be used individually and in concert with other elements to create unique places. In other words, Bismarck, North Dakota has a distinctive feel and character that would be hard to find in Greenwich Village, Manhattan, New York. Place character is a composite, based on the way in which the elements discussed previously are combined. It is hard to anticipate how an individual would feel when immersed in either of the two locations without speaking with them and knowing more about their experiences.

Planners have the difficult task of helping to determine what the character of a place should be in the next 5, 10 or 20 years, a task made difficult because neighborhood character is based very much on individual preferences; and because mobility and natural demographic transitions dictate that the individuals or groups that are involved in planning exercises to define a particular neighborhood character for future years may not be the eventual users of that particular place. With experience, planners and urban designers can identify some enduring themes that create livable places. Lynch, for instance exhorts us to observe people as they navigate their way through large places (Lynch, 1960). Holly Whyte documents human behavior, small microscopic actions[12] that have significant consequences in creating place character. Both these action research pioneers highlight the benefits of careful and systematic observation of human behavior in creating livable places.

While interest and commitment to integrating new technologies came from early adopters in Oak Park and from university faculty, the technology agenda was not driven explicitly by the community or the village's planning staff. Rather, the integration of technology in the project evolved, waxed and waned organically over the lifecycle of the project. However, the successes of the project can be directly attributed to the Oak Park project team's commitment to the ideals of participatory planning – making sure everyone was heard, creating multiple ways for engagement, being transparent about why particular recommendations were made, providing meaningful choices/options that demonstrated the consequences of particular trade-offs, and finally by providing training to staff so that these ideas are seeded within the Village administration. To a large degree, the tools were critical to the success of these participatory planning ideals.

Reflecting back to the work of the SEHPC, it is important to note that the planning described there was conducted at a time when digital technologies were in their infancy, particularly in terms of providing Internet access to ordinary citizens. Between 1995 and 2002, personal computer use increased significantly[13] while capabilities for broad band communication expanded greatly allowing for faster downloads of large files that included maps, renderings, photos, and sometimes, short movie clips. All these capabilities were used effectively in the Oak Park project. In the next chapter, we will learn more about doing participatory planning on a grand scale.

Notes

1. Village of Oak Park: Village Background available at: http://www.oakpark.us/Village_Background/Village_Background.html (Village of Oak Park, 2005).
2. Oak Park Diversity Statement available at: http://www.oakpark.us/Living_In_Oak_Park/Living_In_Oak_Park.html (Village of Oak Park, 2007).
3. Oak Park (1990).
4. Realizing the Vision: the 2040 Regional Framework Plan is available at: http://www.nipc.org/2040/
5. Bentrup (2002a).

6. Ibid.
7. A scaled-down version of the complex is just being completed in 2009, with a municipal subsidy of $4 million after numerous delays.
8. Bentrup (2002b).
9. Al-Kodmany (2000).
10. Lakota Group designs for Harrison Street available at: http://www.thelakotagroup.com/harrisonstreet/
11. UIC plan designs for Harrison Street available at: http://www.oak-park.us/public/pdfs/UIC/10.31.03 planningtogether3.pdf
12. Read Whyte's book *The social life of small urban spaces*, an outcome of his Street Life Project.
13. The Pew Internet and American Life Project, http://www.pewinternet.org/

Chapter 7
Chicagoland's Search for Common Ground

Make no little plans; they have no magic to stir men's blood and probably themselves will not be realized. Make big plans: aim high in hope and work, remembering that a noble, logical diagram once recorded will never die.
Daniel Burnham, architect of the 1909 Chicago Plan

7.1 Introduction

The Northeastern Illinois Planning Commission (NIPC) began work on a long term land use and growth management plan for the Chicago region in 2001. In this chapter, I describe the agency's efforts over a 5-year period to design and implement a collaborative planning process (the Common Ground process) supported by the use of innovative digital technologies. The 2040 Regional Framework plan, formally adopted in 2005 is the culmination of these efforts. As a Chicago resident, I had a front row seat to observe and participate in the Common Ground between 2002 and 2004.

This case study is unique because the impetus for creating the bottom-up planning process came from the agency, driven by a desire to seek meaningful public input into the plan making process. The logistical challenge of the endeavor, specifically the need to meaningfully engage a representative cross section of the region's 8 million residents spread across an area covering 3,750 square miles over a 4-year period also makes it a special case. The use of a range of digital tools and products that created a new standard for participatory planning in the country is yet another reason that this case is unique. Finally, if the city of Chicago and the suburban communities, none of them particularly attuned to a regional way of thinking about the future, could get together to develop a regional framework plan, then this gives cause for hope that regionalism can yet thrive in this country.

L. Ramasubramanian, *Geographic Information Science and Public Participation*, Advances in Geographic Information Science, DOI 10.1007/978-3-540-75401-5_7,

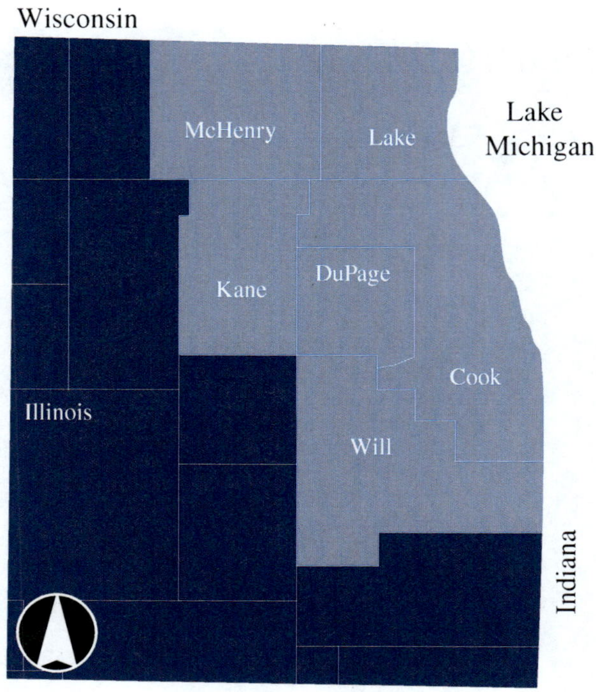

Fig. 7.1 Map of the Chicago six-county region

7.2 A Planned City, A Sprawling Region

Chicago, often labeled the "Windy City", is an American icon. It is one of the nation's busiest transportation hubs, the center of a booming service economy, and an aspiring world city. A vibrant downtown, a range of business, entertainment and shopping districts that are accessible by public transit, and an array of cultural activities have made Chicago a prominent international tourist attraction. The Chicagoland six-county region's population according to the US Census was 8.1 million, a number that is projected to increase by 30% by the year 2040 (Fig. 7.1).

Incorporated as a town as early as 1833, Chicago rose in significance as a regional transportation hub by the 1860s. Chicago's development and physical transformation in the latter half of the nineteenth century is also inextricably linked to the Great Chicago Fire of 1871, which destroyed over 2.7 square miles of downtown property and several hundred lives.

The great losses sustained by the fire influenced architecture, city planning, and building construction. The rebuilding of Chicago followed new guidelines. While many new buildings were constructed, the new zoning codes did not replace the residences that were burned. Downtown business development was encouraged by planners and developers, who were quick to observe the economic advantages associated with the clustering of commercial activities. New construction techniques coupled with the introduction of elevators allowed buildings to be much taller than before. The cityscape that re-emerged after the fire was characterized by a concentration in the downtown area of commercial and business enterprises housed in tall buildings of masonry construction. The area became known as the Loop, and is conveniently served by public transit systems.

Towards the end of the nineteenth century, Chicago hosted the World's Colombian Exposition. This attraction spurred new building construction and attracted young creative people, who made considerable contributions to the development of the city's cultural scene. At the turn of the century, the city was transformed into a gallery of wonderful works of architecture and landscape design. Louis Sullivan's buildings, along with those of other architects (such as Burnham and Root, Holabird and Roche, and William Le Baron Jenney), part of a building tradition known as the Chicago School, dominated the cityscape.

In 1906 Daniel Burnham began work on a master plan for the city. Burnham's *Plan of Chicago* presented a framework for organizing transportation, recreation, and advocated the creation of magnificent public spaces, which were accessible to the general public. Burnham's plan for the "city beautiful" was designed to reduce the concentration of urban dwellings. Perhaps unintentionally, it encouraged urban sprawl. However, even as the city grew and took over the prairie, the areas immediately surrounding the Loop began to decline rapidly, a phenomenon that has only been gradually reversed at the turn of the millennium.

7.3 City Politics and Regional Consequences

Chicago has always been considered a working class city and tensions between labor and management[1] typify Chicago's development almost from the beginning of the city's growth. By the beginning of the First World War, Chicago had evolved into a major industrial/manufacturing center which housed steel and meat packing plants; in turn this created many subsidiary industries. The labor intensive manufacturing activities collectively served as a magnet in attracting a wide range of immigrants to the area including Germans, Irish, Italians, and the Polish. Southern Blacks, migrating north, also settled in Chicago. Living and working conditions for workers and their families were abysmal.

The radical community organizer Saul Alinksy emerged in this milieu, beginning an organizing campaign to turn around declining neighborhoods like the Back-of-the-Yards community that abutted the Chicago stockyards. Alinsky's organizing style was confrontational and appealed directly to the self interest of individuals and marginalized groups. Alinsky was able to mobilize poor people living in miserable conditions, doing menial jobs, essentially those experiencing both material and social deprivation. He persuaded them to work collectively to challenge the establishment, reminding them that they had very little left to lose. He developed and implemented creative and unorthodox tactics to pressure (some might say, to scare) the establishment into giving into the demands of his constituency. The Alinsky organizing model was, and is very effective to accomplish specific short term goals, especially when there is a specific policy or practice to challenge/change (Alinsky, 1971).

The societal shift from a manufacturing economy to a service-based economy did not benefit Chicago. The city of Chicago has shed manufacturing jobs since the 1960s, and the negative consequences of job losses have disproportionately affected the working classes, the poor, and people of color. The people who left the city for the suburbs were mostly white and middle class. As they left, they took with them their tax dollars that funded schools, and other public services. Unfortunately, the politics of Richard J. Daley, the Mayor between 1955 and 1976 were characterized by the manipulation of interest groups based on race and ethnicity. Alkalimat and Gills (1989, p. 20) observed, "Jobs and economic favors were differentially and disproportionately allocated, based upon voting strength, which in turn was based on which ethnic groups were represented". The politics of race became the primary divider between city and suburb. Regional governance did not have much of a chance to thrive, because it would have meant the dilution of black political power within the hyper-segregated city, as much as it would have forced racial equity (integration) on suburban communities.

In addition, Daley's machine style politics placed the city of Chicago and the newly emerging suburbs at odds with one another. The suburbs were dominated by Republicans who considered the city to be a hotbed of dishonesty and political corruption. The city was largely Democratic, and well organized to tap into benefits that flowed directly from the State legislature. Neither party had particular interest in creating legislation nor were they willing to take actions likely to negatively affect their own political power base.

In 1983, a progressive coalition elected Harold Washington as the first African American Mayor of Chicago. Washington's neighborhood development agenda sought to provide a voice for the citizens of Chicago and engage them in democratic decision making. Washington was a reform-oriented politician, who created coalitions across racial lines to further his progressive policy agenda. His first term was marred with conflicts with the city council and his contributions were cut short by his untimely death in office in 1987.

Margaret Weir (2000, p. 147) reports that "when suburban Republican leaders took over the state legislature in 1993, they made their desire to advance a low-tax, anti-Chicago agenda, clear". She draws attention to failure of many city development initiatives, including a plan to create a third regional airport, which she cites as evidence for city-suburb hostility.

Since the mid-late 1990s, Mayor Richard M. Daley made significant investments in city beautification, crime prevention, and school reform efforts that are designed to lure suburbanites back into the city. The city has had considerable success in this regard. In 2004, Millennium Park, a civic space on the shores of Lake Michigan adjoining the loop was unveiled, a visible symbol of the city's resurrection. Chicago made an unsuccessful bid for the chance to host the 2016 Olympic Games, its ambition is another indicator of the city's aspirations to re-enter the pantheon of world cities.

This crash course on Chicago history should make one thing clear – regional governance never took hold in Chicago, despite the creation of two regional planning agencies – the Chicago Area Transportation Study (CATS) founded in 1955 and the Northeastern Illinois Planning Commission (NIPC), founded in 1957. The region continued to grow without a coherent vision. At present, a variety of planning agencies and special purpose boards are responsible for planning.[2]

Transportation issues are managed by the Illinois Department of Transportation (IDOT) and the Regional Transportation Authority (RTA). The RTA provides financial oversight and planning advice for the three public transit operators in northeast Illinois: the Chicago Transit Authority (CTA), Metra commuter rail, and PACE suburban bus.[3] Each of these agencies has their own planning and technical analysis units. CATS published seven long-range plans from its inception in 1955 until it was superseded by the Chicago Metropolitan Agency for Planning in 2006.[4]

As the designated agency responsible for land use planning, NIPC completed its first comprehensive general plan in 1968 which was updated in 1977. In 1992, the agency completed a Strategic Plan for Land Resource Management, a document that drew attention to rampant sprawl in the six-county region; between 1970 and 1990, residential land use increased by 36%, although the population only increased by 4% during that same time period. Subsequently the agency conducted a series of studies focused on land conservation, water quality, and bio-diversity recovery.

Without any enforcement authority to implement its recommendations for land use planning, the agency's real mandate came from compiling socio-demographic data and the development of population projections. These projections became the inputs into the transportation modeling and planning process that happened within CATS. An inter-agency agreement signed in 2000 between the Illinois Department of Transportation (IDOT), the RTA, and CATS formalized this relationship, stipulating that NIPC plans and data were to be the "official" baseline data that were used in the development of the Regional Transportation Plan. The stage was set for the emergence of the Common Ground Process.

7.4 NIPC: An Agency with Leadership and Vision

In 2001, NIPC began work on the 2040 Plan, a necessary task for the development of the Regional Transportation Plan being undertaken by CATS. As the region's Metropolitan Planning Organization (MPO), CATS, specifically the CATS policy committee was required to develop long and medium-range transportation plans in a way that regional investments in transportation infrastructure can be monitored and evaluated. As the agency wrestled with the need to create a participatory process for the regional plan, the need for a new process became quickly apparent – Chicagoland's growing population, its diversity, the limits on land availability, and the need for effective land use-transportation coordination all suggested the need for a more intensive consultation with constituents than ever before.

The architects of the Common Ground Process were NIPC staff, led by its Executive Director, Ron Thomas, and the Common Ground Program Manager, Hubert Morgan. Common Ground's objectives were:

- Prepare and adopt the 2040 Plan based on a publicly created, comprehensive vision for the region's 35 year future.
- Use the best available technology to support a new approach to community-driven regional planning in northeastern Illinois.
- Assist local planning efforts by researching and disseminating best practices, utilizing new technology and facilitating intergovernmental cooperation.
- Strengthen the link between land-use planning and infrastructure investment across the region.

The timing was right for the launch of the Common Ground process. The Federal Highway Works Administration (FHWA) through federal transportation legislation such as the Intermodal Surface Transportation Efficiency Act (ISTEA, 1991) and the Transportation Equity Act for the 21st Century (TEA-21, 1998), had indirectly empowered regional planning agencies by giving them more flexibility to explore transit-oriented, land use planning alternatives. The guidelines handed out to the MPOs require extensive consultation with the public. Furthermore, MPOs are encouraged to develop realistic visualizations of plans and proposals and integrate the use of electronic media in their efforts to engage the public (FHWA, 2007).

Collectively, the attitudes of the federal government can be considered the "new regionalism" (Wheeler, 2002). Wheeler goes on to say that new regionalism is characterized by an emphasis on spatial planning to explicitly address the problems of piece-meal metropolitan growth; advocating a holistic approach that considers environmental, economic and equity goals when undertaking land use and transportation planning. New Regionalism seeks to create a sense of place through careful applications of urban design policies and techniques. In the end, new regionalism takes a normative and activist position *vis-à-vis* growth management.

Ron Thomas and Hubert Morgan, both committed to the ideals of creating a bottom-up planning process, also saw the Common Ground as a way to legitimize NIPC's forecasting work and align NIPC to be more compatible with the work of CATS. Common Ground, thus advocated community-based regionalism, where individual communities maintain their traditional authority, but commit to work collaboratively with one another to accommodate the resolutions arrived at through a regional consensus.

As the manager of the Common Ground process, Hubert Morgan became one of the main public faces of the project as he shepherded the process forward. The core staff for the Common Ground process included ten NIPC planners and three research assistants from the University of Illinois-Chicago. Several technical consultants, an army of volunteers from within NIPC, representatives from universities, community-based organizations, and other agencies participated in Common Ground activities over the 5-year period.

7.5 Key Elements in the Common Ground Process

The overarching goal of the Common Ground (CG) process[5] was to develop a comprehensive vision of the region in 2040. The project achieved this goal by including the following activities:

7.5.1 Leadership Workshops (Fall 2001)

CG staff organized a series of meetings where business and community leaders could come together to talk about topics of concern in their neighborhoods and municipalities. The leadership workshops were held in 12 locations across the six-county region. These meetings were of great importance to CG staff. During these conversations, staff learned about issues and concerns that were foremost on the minds of participants. Yet, they also gathered information about other practical matters, including how to expand their outreach efforts. The leadership workshops and other participatory activities eventually resulted in 52 goals statements.

7.5.2 Regional Forum (Fall 2001)

The Regional Forum was a town hall meeting, where people from different walks of life gathered in one place. While the leadership workshops were held in different regional locations and brought together people from specific sub-regions, the discussions tended to emphasize issues pertinent to that sub-region (e.g., farming, or water supply or jobs). On the other hand, the regional forum brought together 900 (!) people, who represented the diversity of the region – old and young, of different social classes, races and ethnicities, with different occupations and concerns. In particular, the regional forum closed the city-suburb divide. Participants spent time identifying what they liked and valued about their region. In addition, they also prioritized the issues and challenges identified at the leadership workshops.

7.5.3 Working Groups (Fall 2001 to Summer 2002)

Over 250 people, who participated in the regional forum and earlier meetings, signed up to do some intensive focused work to develop regional goals and a new form of regional governance. 20 groups formed around 5 themes and 4 geographical areas. Eventually, the 52 goal statements were organized into five themes: livable communities, diversity of people, healthy natural environments, global competitiveness, and collaborative governance. The members worked over an 8-month period, meeting face-to-face and electronically. Each working group developed a mission statement, conducted background research and conducted a conventional Strengths,

Weakness, Opportunities, and Threats Analysis (SWOT analysis) to create a robust understanding about the issues facing the region.

7.5.4 Youth Forum (Spring 2002)

Although the participation of youth aged 14–18 years at the regional forum was approximately 10%, CG staff felt that they had not stayed involved, because many of the activities that followed the Regional Forum were evening activities, not conducive to youth participation. Thus, in April 2002, a special youth forum was organized in conjunction with the National American Planning Association conference. That Saturday morning, about 90 young people from around the region participated in a discussion about the future of the region.

As a volunteer facilitator, I found the process to be engaging and rewarding. There was a regional map on the table and pictures of iconic regional locations. The youth were asked to place the pictures at the appropriate location on the map. This activity created a lot of discussion – and information sharing. Subsequently, the youth spoke about their hopes and fears for the future of the region, conducting their own SWOT analysis. Their main concerns were about the environment, alternative energy sources, and alternative forms of transportation. Looking back, it seems the youth were very prescient.

7.5.5 Goals Writing Workshops (June 2002)

The 20 working groups reassembled as 5 groups to discuss and resolve differences of opinion to finalize the goals statements. The project leaders report that "using working groups to draft the goals themselves – rather than leaving the responsibility to CG staff – has been an integral part of CG and the process of community-driven regional planning" (2040 Regional Framework Plan, p. 94).

7.5.6 Goals Review Workshops (Fall 2002 to Spring 2003)

Additional feedback was solicited from traditionally under-represented populations. Participants in these meetings reviewed the work done by the goals-writing teams and made comments and suggestions. Elected officials also reviewed the goals statements in separate workshops. A special workshop was held to address the needs of farmers in one county.

7.5.7 Commission Endorsement (March 2003)

NIPC formally endorsed the 52 goals and the set of 5 core themes (See Fig. 7.2).

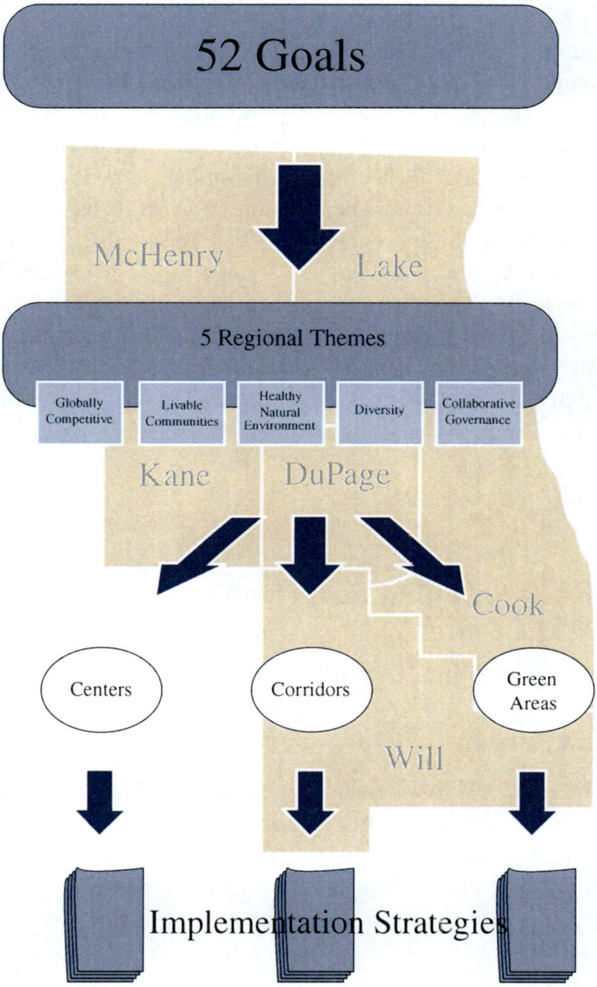

Fig. 7.2 Goals, regional themes, and implementation strategies

7.6 An Integrated Land Use/Transportation Plan

Once the goals and regional themes had been endorsed, it was time to create the integrated land use/transportation plan. This task, too, was conducted through workshops, where the focus was on mapping the region using a custom-built tool called Paint-the-Region. The tool will be discussed in the next section. In 2004, thirteen separate workshops were conducted. Participant numbers and demographics varied by workshop. In general, each workshop included a room with 5–8 tables, each table accommodating 8–10 participants. The groups worked to assign activities in specific locations on the map. They were supported by a technical facilitator and a discussion facilitator.

The participants were able to digitally add or expand nodes and activity, expand existing linkages or create new transportation corridors, and identify green areas that needed to be preserved. Participants were able to review existing data about the area they were working on, including data about population, land use, transportation infrastructure, environmental features (rivers, wetlands, etc.). Having participated in some of these workshops, I can report that each mapping task was preceded and followed by discussions about how the mapping choices related to the overall regional themes. At the end of the each workshop, CG staff was able to create a synthesis map that essentially "quilted" together the work that occurred at the different tables. Eventually, they created a regional composite map that integrated the quilted maps from each of the 13 workshops. The result was the Regional Framework Map that identified centers, corridors, and green areas.

7.7 Digital Tools for Public Participation

7.7.1 Facilitation Tools

7.7.1.1 Key Pad Polling

At the regional forum and other large group events, digital key pad polling was used to engage participants. Participants expressed preferences, priorities, and opinions about a variety of topics, anonymously. The main payback of key pad polling is the immediate feedback that can be generated about how the group feels about any particular issue. This is useful for the participants as well as the organizers.

7.7.1.2 Web Council

Working groups communicated electronically via WebCouncil, a collaboration platform that allows spatially dispersed groups to share ideas and to coordinate actions.

7.7.1.3 AmericaSpeaks

NIPC used the assistance of a nonprofit group called America Speaks to facilitate and manage the activities of the Regional Forum described in Sect. 7.5. AmericaSpeaks is a facilitated discussion process aided by networked computers managed by technical facilitators. The goal of the AmericaSpeaks process is to ensure that people are engaged respectfully – everyone must be heard and feel that they were heard, that their time was not wasted, that their viewpoints were taken seriously. In the AmericaSpeaks process, the discussions are synthesized and shared with the participants almost immediately, thereby helping to focus the discussions as the meeting progresses. The method was used effectively in the listening sessions convened in New York city, where 5,000 people came together to plan the rebuilding of lower Manhattan after September 11th and in the discussions about how to rebuild New Orleans after Hurricane Katrina.

7.7.2 GIS Tool: Paint-the-Region

Paint-the-Region (PTR) is customized software that is based on INDEX, a suite of GIS tools developed and supported by Criterion, a national urban planning consulting company. Paint-the-Region allows end users to digitally assign land use typologies onto parcels on the map by selecting from a palette of graduated colors and placing them on the map canvas. The technique makes it easy for non-technical users to make decisions about land use choices with a minimum of effort. In addition, end users can draw growth boundaries or new transportation corridors. Paint-the-Region runs as an ArcView extension and can be deployed on standalone computers or accessed through a web browser. The benefit of having the visual capabilities linked to a robust GIS is the ability to be able to compute the impacts, including the number of new households, the increase in population, the new jobs that can be generated, and the acreage of land that will be lost to development.

The Paint-the-Region palettes also provide visual cues about place character – providing the end user with impressions that more appropriately describe centers (metro area, commercial center, town center), green areas (agriculture, open space, water, trails, and so on), or transportation corridors (heavy rail, expressway, bus rapid transit, and so on). The emphasis is in creating realistic representations of places so that end users know the consequences of designating a parcel or a group of parcels as a particular land use.

7.8 Final Outcomes

The 2040 Plan took an integrated approach to dealing with the issue of land use and transportation coordination and the process reinforced the regional issues and common shared future, rather than the conventional political jurisdictional boundaries that create the tensions between the city and the suburbs. The decision of the CG staff to approach the plan with the three-pronged framework of centers, corridors, and green areas provided a simple way to link the 52 goals, through the 5 themes to specific implementation strategies (See Fig. 7.3). The plan was formally approved on June 29, 2005.

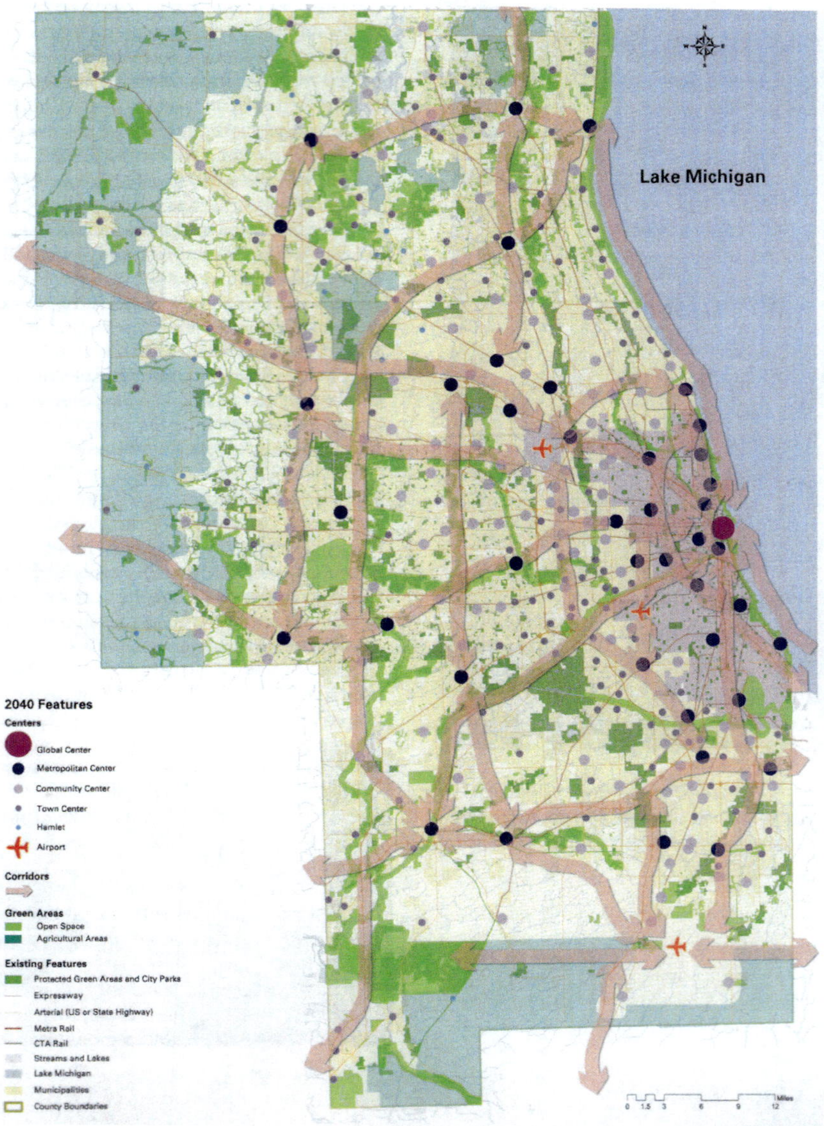

Fig. 7.3 Regional framework map

7.9 A New Regional Planning Agency

NIPC and CATS were merged to form a new agency called the Chicago Metropolitan Agency for Planning (CMAP), finally integrating planning for land use and transportation in Northeastern Illinois. An MPO Policy Committee will take

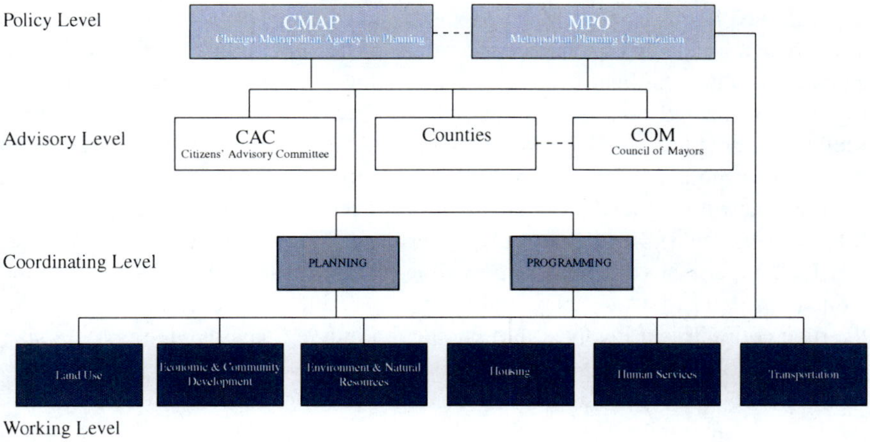

Source: Chicago Metropolitan Agency for Planning

Fig. 7.4 CMAP organizational chart

responsibility for MPO functions. CMAP's mission includes: research and analysis, land use, transportation, economic and community development, environment and natural resources, housing and human services. The agency's by laws were revised in November 2007. It is much too soon to tell whether the new super agency will be effective in sustaining a regional vision for Chicagoland (Fig. 7.4).

The 2040 Regional Framework Plan has now morphed into the *GO TO 2040* Campaign to develop and implement strategies for addressing the consequences of the projected growth of the region by 2.8 million residents and 1.8 million jobs by 2040. The campaign has a prominent place on the agency's website, and it appears that the agency is continuing with the same originality that set the CG process apart. New technologies and modalities (video, online surveys, a digital photo contest, and blogs) are some of the innovations that are being included in this phase of the regional planning process.

7.10 A Comment on Regional Planning

Regional planning appears to be enjoying a renaissance. The idea that some problems are better managed and solved by thinking regionally has gained some ground (Albrechts, Healy, & Kunzmann, 2003; Shaw & Sykes, 2005). Chief among these problems is unplanned growth or sprawl. Managing sprawl, while supporting planned growth requires regional strategies, agreements, and considerable coordination in infrastructure investments, as well as careful assessments of negative externalities such as increased traffic, air pollution, and noise in order to create a sense of place (Yaro & Hiss, 1996; Calthorpe & Fulton, 2001).

Regional planning work of the sort undertaken through the Common Ground process is often highly technical and voluminous. The obligations of regional planning agencies to integrate land use and transportation planning with the goal of reducing congestion and providing increased mobility requires participants to become proficient in reading and interpreting the language of land use planners, civil engineers, traffic modelers, and economists. More importantly, participants must become comfortable with the idea of making complex decisions with imperfect information (Stephenson, 1998). Inevitably, a credible process becomes an educational process, in which experts are involved in providing testimony and advice to non-technical citizens in order that they may make reasonably informed decisions. As a result, the time commitment involved in participating in a regional planning process is far greater than a local project planning effort.

Ultimately, the NIPC Common Ground process engaged over 4,000 participants (residents, community leaders, elected officials) in a workshop process in order to establish a shared vision for the future and a process to achieve those goals. The sheer scale of this participatory planning endeavor necessitated the extensive use of geo-spatial technologies and e-participation methods. Specifically, the CG process included innovative new ideas of process by (i) integrating local land use planning and regional transportation planning; (ii) creating many opportunities for small group meetings in many communities across the region, including targeted involvement of youth, minorities, and non-English speakers; (iii) and returning to these groups to show them planning analyses at different stages, and conducting focus groups to solicit feedback.

Without a federal or even a state-wide mandate to support regional planning, the work of regional planning agencies relies on building strong coalitions among diverse stakeholders. This case study affirms the observations made by Innes & Booher (2004) that the processes of regional spatial planning require the mobilization of networks of actors at the intra-regional levels.

Notes

1. For example, the Haymarket Massacre and the Pullman company workers strikes and the race riots of 1919.
2. This fragmentation is not unique to Chicago. Many major US cities, including New York experience the same dilemmas.
3. Regional Transportation Authority, see http://rtachicago.com
4. Historical Regional Transportation Plans for Northeastern Illinois, see http://www.cmap.illinois.gov/sp2030/historic_plans.aspx
5. Many individuals were involved in the Common Ground process, some as staff, others and volunteers and friends. However, the leadership exerted by the agency's senior management was central to the project's success.

Chapter 8
Evaluation

8.1 Introduction

As previously observed, much of contemporary planning practice is subsumed within a rational planning paradigm. This worldview assumes a teleological and pragmatic problem solving approach that begins with the establishment of goals and concludes with the implementation of the plan. But, how we do know we've succeeded. That's where evaluation comes in. In everyday terms, evaluation consists of systematic and careful assessments of individuals, projects, programs and/or policies. As individuals, depending on our situational context, we may sometimes experience evaluation as a positive and supportive set of analyses that enable us to do better and achieve our personal and organizational goals; however, for most of us, evaluations are associated with external scrutiny where "outsiders" look for things that went wrong in order to assign blame. In an organizational context, evaluations are often used to limit or withdraw funding, modify policies, and in extreme cases, completely shut down activities, projects, programs, or even entire organizations.

Evaluation relies on quantitative measures and scientific methods. Yet, social scientists now acknowledge that a variety of qualitative methods and techniques can be used to conduct evaluations – a range of methods, including individual interviews, focus groups, place and people-centered behavior mapping, and the analysis of textual, graphical, and verbal narratives are part of the repertoire of evaluation researchers.

Empowerment[1] has been one of the overt goals of the PPGIS movement, in the fifteen-or-so years that PPGIS activities have flourished in the United States. However, the empowering qualities of PPGIS work are difficult to evaluate, in large part, because PPGIS activities are often embedded within larger initiatives with broader organizational goals. In addition, it is difficult to document intangible benefits that accrue from a particular project and develop a causal linkage with a specific PPGIS activity. In Chap. 3, I laid out a framework for understanding PPGIS in this way. We learned that PPGIS cannot be understood without taking into consideration the goals (purpose) of public participation; the characteristics of the participants (demographics, cultural characteristics), the methods of community engagement; how the participatory process was designed and managed, the tools and techniques

L. Ramasubramanian, *Geographic Information Science and Public Participation*, Advances in Geographic Information Science, DOI 10.1007/978-3-540-75401-5_8,

that were used; the data and information used as inputs and generated as outputs, and the project timeline.

PPGIS evaluation is a messy enterprise. Should we subject PPGIS case studies to a uniform evaluation framework? Comparing PPGIS activities in small fishing villages in Indonesia with the work of community organizations in New York City seems akin to comparing apples with oranges. If the situational contexts and goals of each PPGIS project are so unique and special, then is there any benefit in attempting to draw generalizable conclusions about what constitutes "successful" or "unsuccessful" PPGIS projects?

Despite these concerns, I propose that it is reasonable, in fact, necessary to evaluate PPGIS activities. When considering particular social issues, improving the health status of indigenous peoples for example, there may be value in comparing information and evidence cross-culturally using a rigorous framework – lessons from PPGIS experiences among communities of color in New Zealand may well benefit PPGIS work among communities of color in the United States. Likewise, most researchers are aware that there are islands of social and material deprivation[2] enmeshed among the largely affluent socio-economic conditions in any society. Thus, PPGIS work with homeless youth in Boston, USA may have many parallels in work with homeless youth in Sydney, Australia, even if the USA and Australia are both relatively privileged and affluent societies.

However, we need a stable meta-evaluation framework to organize our evaluation attempts. The framework must be useful enough to help organize and compare different types of projects, and allow us to account for the high variability we are likely to encounter, but at the same time, the framework must provide enough structure so that we can draw together seemingly disparate projects and compare and contrast them effectively. In this chapter, I demonstrate that such a framework can be useful and use it to evaluate the three case studies I presented in Chaps. 5, 6, and 7. The principles of participatory research discussed in Sect. 8.2 guide the development of the evaluation framework.

8.2 Participatory Research and Evaluation

Assessments and evaluations are always purposeful; therefore the selection of assessment and evaluation methods must be appropriate to the task at hand. Good evaluation research emphasizes rigor, integrity, transparency, and the systematic gathering of evidence to support conclusions. Evaluators are obligated to analyze and present all data, including data that does not support their own hypotheses/expectations (e.g., Werner, 2004; Denzin & Lincoln, 2005).

Advocates of community-centered planning have consistently argued that citizens, especially those affected by the planning processes, must be actively involved in the evaluation. In the international contexts, where large scale human and social interventions are carried out through the efforts of the World Bank and other aid agencies, the notion of engaging intended beneficiaries in the evaluation of projects has become firmly established. Robert Chambers is often credited with validating the role of participatory assessment and evaluation. In his 1995 book, *Whose reality counts? Putting the last first*, he vociferously argues that actively engaging the intended "beneficiaries" in the evaluation of any planning initiative is essential to understand and measure the success of failure of any initiative. Since his work was focused on rural development, Chambers criticizes outsiders for having an "urban bias". He harshly concludes that academic/professional evaluations of rural situations are often framed in negative terms because of these outsiders' inability to perceive and understand the creativity and innovations happening on the ground.

The initial focus of his book is about using appraisal (in other words, a needs assessment or a formative evaluation) to assist in on-going planning and decision-making. He further argues that participatory appraisal is neither unscientific nor time consuming, arguing for simple methods (such as participant observation, focus groups, and the collection of narrative accounts) to gather evidence from the community. Participatory Rural Appraisal (PRA) is now part of the repertoire of evaluators proceeding to rural communities world-wide. PPGIS practitioners, particularly those working in countries outside of the USA have become quite devoted to the methods advocated by Chambers (Rambaldi et al., 2006).

PRA makes use of visual and graphical representations as communication aids in workshop settings. For Chambers and his colleagues, participants must be actively involved in creating the representations, thus shaping the narratives. PRA advocates argue that while verbal discourse can sometimes marginalize vulnerable sub-populations (e.g., women), creating a community map that describes assets, perceptions, and routes can reveal important highly localized information to planners. For instance, in rural communities, where women and children bear primary responsibility for gathering food, fuel, and water, it is important to understand the lived experiences of women and children, including the hardships they experience in fulfilling mundane chores. These hardships are sometimes elegantly communicated on maps and drawings and allow the marginalized women to speak more "loudly" and "clearly" than in a conventional community forum, where they may not have the opportunity to speak. In this example, the act of drawing the maps, and the stories narrated through the maps are empowering.

Chambers' ideas have helped to transform the design, management, and evaluation of development projects and programs. However, there is some concern among researchers and practitioners that PRA has become formulaic and routinized (e.g., Henkel & Stirrat, 2001). Chambers, by publishing *Participatory workshops: A sourcebook of 21 sets of ideas and activities* (2002), perpetuates the myth of a highly generalizable approach to participatory work that can be easily learned and implemented.

Critics of participatory rural appraisal challenge PRA's simplistic understandings of "community" that ignores or glosses over deeply embedded power differentials within the community. For example, women who are silent in conventional community meetings, are hardly likely to become animated participants, simply because the nature of participatory activities now includes the creation of perceptual maps. They may still feel obligated to toe the "party line", i.e., those values that are espoused by village elders, their husbands, fathers, or brothers, or the views of the state or local government that is eager to receive development aid. Participatory rural appraisal imposes new burdens on already vulnerable people and communities, – engagement comes at a price – of time invested, labor expended, and ultimately, with the risk of political cooptation (Cooke & Kothari, 2001).

Outside international development circles, the term "participatory research" has become *en vogue* to signify a learning-oriented approach that can simultaneously achieve programmatic goals while creating opportunities for reflection and assessment. According to Hall (1993: xiv), participatory research combines "research, education, and action". By explicitly linking these three activities, participatory research establishes clear linkages between knowledge generation and knowledge utilization. In addition, participatory research places the traditional research process (problem formulation, identification of an research design, determination of methods of data collection and analysis, and the elucidation and refinement of research findings) in the hands of those who are usually the "researched". According to Park (1993, p. 3), "participatory research captures the ideal of goal-oriented, experiential learning, and transformative pedagogy", drawing from the theories of Freire and Schön. The long term goal for a participatory research endeavor is to empower people psychologically and politically to effect social change.

Critics of participatory research are usually opposed to the use of the term *research*; they note that careful scientific inquiry should not be confused with community building activities. Critics also make the claim that lay persons lack the ability to carry out complex research-oriented tasks like problem identification, development of data collection instruments, data collection and data analysis. These criticisms appear to stem from ideological positions (world views) and hence there seems to be little point in countering them. For my part, I argue that it is possible to design and implement participatory strategies in a systematic way so as we can learn from the implementation process. Ultimately good participatory research can establish a balance between the goal of rigorous scientific research and conducting research about issues that have practical relevance in naturalistic settings.

Regardless of the terminologies used, contemporary social scientists believe that evaluations should emphasize respect for both people and institutions participating in the evaluation process. Evaluation can be formative (with a goal of assisting participants in the refinement and development of a better process/problem solving strategy) or summative (with a goal of assessing impacts and outcomes of a particular program or programs) (Werner, 2004). Thus, participatory research can help remove the stigma of an "external evaluation" and create meaningful opportunities for participation of end users.

8.3 PPGIS Evaluation

An ideal PPGIS project is a participatory planning project that is supported with digital technologies. At a *minimum*, it should include the following ingredients:

- develop the capacity of the participants to organize, analyze, and discuss planning concepts to the level required by the particular endeavor they are involved in;
- engage participants in every aspect of the planning process, that is, in the framing the project goals, the methods that are selected to examine and investigate these goals, in project implementation, and assessment;
- develop techniques to carefully incorporate participants' views and participant-generated data into formal planning processes; and,
- provide clear and transparent strategies for data generated from the project to be available to the participants.

Akin to Arnstein's (1969) ladder of political participation, and Voogd and Woltjer's (1999) guidelines for ethical planning, the definition of the ideal PPGIS project as stated above, establishes a goal that all projects/programs can aspire to meet.

The purpose of this chapter is to capture the unique as well as the ubiquitous ways in which PPGIS-based advocacy work has transformed the day-to-day planning practice in the USA. By examining how neighborhood and community institutions have altered or changed their established practices because of their exposure to, and use of geo-spatial technologies, I seek to highlight both positive and negative impacts of PPGIS adoption and use. Furthermore, by examining the extent to which PPGIS practices are successfully established within the day-to-day vernacular of institutionalized planning practice, I hope to stimulate a more robust debate about the best ways to better embed the use of participatory planning methods and geo-spatial technologies within planning and decision-making processes.

8.4 Evaluation Framework

The subject of PPGIS evaluation has not been tackled extensively in the literature. There is some literature on how to evaluate PPGIS tools (e.g., Haklay & Tobón, 2003), but the evaluation of PPGIS tools is not the focus of this chapter. Barndt (2002), discussing evaluation from an applied perspective, places a lot of emphasis on understanding the value of PPGIS initiatives. He rejects an academic approach to PPGIS, arguing that the quality of PPGIS work should be judged by whether it generates results for the client (community). While this an excellent starting point, tying PPGIS effectiveness to new data generation seems rather limiting. Seeking to create a more robust but manageable evaluation framework I consider three elements:

First, I ask, what are the characteristics of the *process design* that was used to introduce geo-spatial technologies within a specific organizational or institutional context? In other words, how was the program planned and developed? This component must include a discussion about project/program goals, participants, methods of engagement, selection of tools, and the management of data and information.

Second, I ask, what is the range of *short term outcomes* that emerged immediately after the program or PPGIS implementation effort concluded? Were these gains and losses planned for/anticipated or were there unintended consequences? What happened during or immediately after the project was concluded. Were project/program goals achieved? Did the challenges bring about the results that were sought?

Finally, I ask, what are the *long term impacts* of these efforts after some time has elapsed? Did everything return to a "business as usual" scenario? Did the PPGIS activities result in observable and enduring changes in planning practice?

8.4.1 Process Design

Planning that precedes introduction of geo-spatial technologies to a community is critical to the success or failure of the implementation, an observation extensively supported by researchers (Rogers, 1983; Onsrud & Pinto, 1993; Obermeyer & Pinto, 2008; Campbell & Masser, 1995; Huxhold & Levinsohn, 1995; Harris & Weiner, 1998 among others). Non-technical factors including the presence of GIS champions, skills and motivation of users, technological congruence with organizational needs, leadership support for information-driven solutions, and political imperatives all affect implementation efforts. The challenges are far greater for PPGIS adoption and use, because PPGIS practice includes the additional obligation/burden to include credible participatory processes within the implementation effort. Thus a PPGIS implementation must be preceded by careful and conscious attention to process, in which the roles and mandates of participants are clearly defined. In the United States, good PPGIS practice is modeled after good community development practice, wherein PPGIS advocates can serve as community organizers (Rivera & Erlich, 1992). In addition, PPGIS advocates concerned about long term sustainability, will attempt to link and integrate their work to on-going planning initiatives that are underway.

8.4.2 Short Term Outcomes

The introduction of new technologies and innovations often promises efficiencies – in terms of use of staff time and resources. More significantly, GIS has been most productive in routine task automation, a feature used effectively in the day-to-day business of planning (Huxhold, 1991; Ramasubramanian, 1999). Evidence of these efficiencies can be observed in customized map production using data that has been assembled and organized from data providers. These efficiencies are increased with the advent of the internet, a transformation that has moved PPGIS away from individual desktops to the interactive public realm. Examples of such internet-based data providers include the US Census American FactFinder (for socio-demographic information) and DataPlace (for housing and community development information). Localized community-based data providers abound, although data quality is variable. Even if one assumes that PPGIS advocates may be able to achieve efficiencies in some routine tasks, benefits are gained only if they redeploy time and resources to meet other needs (like reaching under-served populations or conducting more thorough analyses). Information dissemination is another short term goal that most PPGIS advocates should seek, specifically to get their issue heard by a wider audience; to engage multiple publics; to foster conversations and debates about the issues. A third short term outcome would be an immediate successful resolution of a problem or controversy. In policy controversies such as the need to achieve social equity goals, data-driven analyses can result in "quick wins". In this context, the creative use of digital technologies to support multiple or alternative representations of the issues would be a short term impact. Negative impacts too must be considered in analyzing short term outcomes. Project cost overruns, technical problems, staff burnout, exacerbation of existing tensions within communities are examples of likely short term outcomes that PPGIS advocates must strive to avoid.

8.4.3 Long Term Impacts

Essentially, long term impacts can be grouped into two categories – impacts/changes to process (the ways in which planning takes place), and impacts/changes in policies and programs. These long term impacts are those gains that inspired the initiative in the first place, but may not have been accomplished when the initiative was concluded. Thus, if the goal was to create a more transparent and inclusive planning process, then a long term impact would be the creation of mechanisms and processes that support such inclusive planning. Examples of such impacts include the creation of community councils to monitor planning initiatives or the inclusion of a review/comment phase in a process that formerly did not include that component. Likewise, long term impacts are the establishment of policies and programs that were deemed desirable goals when participatory initiatives were initiated. Examples can include changes in policy to achieve social equity/social justice goals, or the creation and support of programs to monitor such goals. With this framework, a review and analysis of three case studies will provide the much needed context to anchor discussion and synthesis.

8.5 Summary of Case Studies

Three case studies, discussed in depth in Chaps. 5, 6, and 7 are briefly summarized here. Table 8.1, below provides a quick overview. In the previous chapters, I described the case study in some detail. The purpose of devoting an entire chapter for each case study was to provide rich contextual information, identify project goals, describe the different individual and institutional actors, elaborate on the methods used to facilitate participatory planning, specifically focusing on how geo-spatial tools were integrated into the planning activities. The individual chapters also described an approximate project timeline, including the kinds of data that were assembled and the new information and insights that resulted as result of the process.

In each case study, geo-spatial technologies were adopted and used to achieve a variety of planning goals. These cases were chosen strategically to illustrate and

Table 8.1 Overview of case studies

	South end/Boston	Village of Oak Park	Chicago Region
Principal players	South end housing and planning coalition; Boston Redevelopment Authority; south end residents	University-based research/implementation team comprising of faculty and students; Village of Oak Park staff; elected officials; Oak Park residents	Northeastern Illinois Planning Commission staff and volunteers; elected officials; community groups; Chicagoland residents
Project time frame	1996–1997	2002–2003	2002–2005
Issue/focus	Neighborhood planning; preserving affordable housing; ensuring economic development that avoids jobs-skills mismatch	Retail district planning; balancing need to preserve community character while focusing on economic revitalization; planning for future growth; transparent and accountable planning	Regional planning; sustainable development that balancing the needs of present and future generations; quality of life issues
Scale of intervention	Neighborhood	City	Entire region

explicate the usefulness of the framework discussed in Sect. 8.4. Since the framework requires that attention be paid to the planning process, and an assessment of short and long term impacts, I selected cases where I have extensive in-depth knowledge about the context of the case and personal familiarity with many of the activities undertaken to achieve project goals, as a participant-observer, or as an architect engaged in implementing the PPGIS initiative. Yet, these cases include many activities that fall within the purview of local government planning in the United States and therefore contain some generalizable lessons.

The South End Planning and Housing Coalition case study (Chap. 5) exemplifies the types of community mapping and analysis activities that were carried out by an organized and politically savvy coalition representing several community groups in Boston's South End. Despite the fact that they relied on a planning consultant to produce the maps, the coalition members directed the work in a very hands-on way. While it was a small group that actually participated in coalition meetings, they were true representatives of the community, having lived and worked in the neighborhood and their long individual histories of community activism in addressing quality of life issues in the South End provided them with community credibility.

The Village of Oak Park case study, described in Chap. 6 is an example of a university-community partnership project, a popular model of PPGIS dissemination into community settings. Oak Park was and is at the forefront of progressive politics. The citizens and elected officials of the Village of Oak Park espoused policies and practices that valued social and economic diversity as well as broad participation in civic decision-making. The university's investments in supporting PPGIS work were focused on developing frameworks for communication, where technologies were used creatively to explain trade-offs and choices that were being made as the village residents planned the future of two under-performing retail business districts.

In Chap. 7, I describe the Common Ground initiative undertaken by The Northeastern Illinois Planning Commission. Although, Common Ground appears to be a very top-down planning process at first glance, this initiative was a large scale participatory visioning exercise that engaged hundreds of Chicagoland residents in creating a regional plan. Citizens collaboratively generated 52 goal statements that were organized into five themes: livable communities, diversity of people, healthy natural environments, global competitiveness, and collaborative governance. The sheer scale of this participatory planning endeavor necessitated the extensive use of geo-spatial technologies and e-participation methods. This project showcased the complexities of scaling-up PPGIS activities, typically limited to local community activities.

8.6 Applying the Framework to the Cases

In Chap. 3, I stated that the design of a "good" participatory planning process must consider how to involve citizens every step of the way. Any community-oriented activity should begin with the articulation of goals. Ideally, these goals should be determined in close consultation with a wide cross-section of the community. However, in practice, things work a little differently.

Goals are often articulated from outside the community. For instance, a philanthropic organization may release a report that makes a case for additional investment in a particular urban issue such as the creation of affordable housing or in early childhood education. An op-ed in a local newspaper may highlight social problems in a particular neighborhood. New targeted funding opportunities may be announced, prompting local civic leaders to investigate ways to access these resources. Each of these external drivers can result in the articulation of goals for a particular neighborhood/community well before the first citizen is consulted.

Designers of participatory planning processes should not worry too much about how and when these goals are articulated because it is almost impossible to determine the exact point when there is a community consensus that "something needs to happen". However, most organizers know that community consensus sometimes emerges when the community is under great stress. In a crisis situation that needs immediate attention, (e.g., news that a local store is closing, or that a child was hurt in handgun violence), even community residents who are typically apathetic or uninvolved will come together even if it is for a brief moment. Candlelight vigils and protests are not uncommon at this stage. These moments of crisis typically force individuals to move away from a position of status-quo to try something different (Whyte, 1991). In other words, participation occurs, because non-participation (avoidance) ceases to be a realistic option.

For community organizers like Alinksy (1971), these moments of crisis are crucial to build community support for a particular agenda. An organizer using Alinsky's tactics will try to channel the nervous energy of the mobilized masses and focus it on a specific target. Alinsky's tactics are inherently confrontational and oppositional; they are wonderful to begin the process of critical thinking – why are we in this situation? Why it is our community has less resources and investments than that community over there? And so on. However, Alinsky's approaches are not useful for institution-building. And institution-building is essential to secure the gains made by community organizing.

I propose that in each of the three case studies, the designers of the participatory planning activities did not unduly concern themselves with the question of "who came up with these goals?" In contrast, they concerned themselves with the question – "are these the right goals for us to work on, at this time?" Their decisions were informed by their experiences living and working in the community. In the case of Boston's South End Community Housing and Planning Coalition (SEHPC), for example, the Executive Director had been monitoring a variety of issues that were of concern to South Enders. He helped to establish and staff a neighborhood planning coalition. This core group of participants (some were staff of neighborhood planning agencies or staffers who worked for elected officials) met regularly, usually during evenings and weekends. The focus on institution building was enabled by creating a small neighborhood "kitchen cabinet" that proactively monitored development and changes in the neighborhoods, rather than reacting to issues as they unfolded. The group of 20–25 people included working people, retired people, men and women, old and young. These people advocated on behalf of other neighborhood residents who were unlikely to maintain regular attendance at community meetings. Some of these participants viewed coalition meetings as part of their job description, although meetings occurred during evenings and weekends. Others volunteered their time and expertise because of their deep connections and commitment to the neighborhood.

SEPHC was successful because it had a community-based meeting place and a staff member (actually the Executive Director) who could be counted on to send out the e-mails or phone calls, and undertake the basic organizational activities necessary to make community meetings happen. The relative stability provided by the Executive Director and the community people who were neighborhood residents' engendered trust, both among those neighborhood residents who did not regularly attend planning meetings, as well as the official planners responsible for community outreach.

The results of the GIS mapping and analysis activities discussed in Chap. 5 were centered around educating and mobilizing this core stakeholder group, that in turn was able to educate and energize different constituencies – thus information reached the historic preservationists, the affordable housing advocates, the urban gardeners, and advocates for health and well-being. Each of these groups in turn devised focused organizing and outreach campaigns to raise awareness among their constituencies. This de-centralized model of community outreach is an essential part of successful participatory process design.

The design of participatory planning processes is critical because well-designed processes engender trust (Ramasubramanian, 1999; Witten et al., 2000). Planners working in communities like Oak Park (where there is a high degree of civic engagement measured by indicators like voter turnout in local elections and attendance at PTA meetings) find that invitations to get involved in planning initiatives often attract well established community stakeholders most likely to hold entrenched policy positions. The "vocal minority", as these stakeholders are sometimes derisively called, engage in community decision-making processes in order to further a specific policy agenda, thereby resisting efforts to create a consensual approach to plan-making (Innes and Booher, 2004).

The partnership between the university (UIC) and the community (Oak Park) described in Chap. 6 was negotiated between university faculty, the elected officials in the Village, the Village manager, and his planning staff. Citizens were not directly involved in these negotiations. Thus, when the project began, the university could not claim complete neutrality. However, the UIC planners began to reconstitute the framework for civic engagement anew. The planning process began with a simple premise that the first few meetings were simply to listen, and to learn. Furthermore, the goal was to listen and hear from many people, and people who usually did not get involved. Therefore, the UIC planners decided to take primary responsibility for advertising for the first community meeting. Rather than relying completely on the Village's mailing lists, the UIC planners began to create different types of meeting opportunities.

The first kick-off meeting for the project was heavily publicized via conventional door knocking, in print, and through email alerts. On the day of the meeting, the Village manager and his staff were pleasantly surprised that over 75 people turned out on a regular week night to talk about planning and design issues that affected only a small part of the Village. A typical planning meeting in Oak Park would probably attract about half that number, usually the "regulars". By simply beginning afresh, the planning team was able to energize the process. The new beginning also helped create more trust in what eventually came to be called the UIC process.

In each of the three cases, the maintenance of rich contacts with the community relied on key individuals (the Executive Director of SEPHC, the Project Managers of the UIC process, and the Director of Public Outreach for the NIPC Common Ground process). These individuals were simply very accessible, and open minded in discussing difficult issues. While it is not impossible for an outsider to achieve this level of familiarity and intimacy with a community, having an engaged local community leader without a personal agenda can be a key ingredient to ensure that the process design is tailored appropriately to community needs.

So, in the case of the UIC process, we have to ask: what happened when the cameras left, when the students and faculty moved on to other projects and lives returned to conventional routines? It is heartening to note that in February 2006, about 2 years after the Planning Together project concluded, the Village of Oak Park developed Guidelines and Procedures for Participatory Planning that govern the development or re-development of any Village-owned land (Village of Oak Park, Board of Trustees Policy, 2006). The guidelines state:

> the purpose of creating the public participatory planning guidelines is to ensure that each village-owned property being considered for development/ redevelopment is reviewed in a consistent and open manner...

The guidelines emphasize open communication and the need to raise awareness about planning issues in the Village across a wide swath of the public and the need to provide multiple opportunities for review and comment. The Planning Together process showed elected officials and planning staff that most citizens understood the need to make trade-offs and were able to balance their interests and commitments to maintaining community character with the needs of growth and economic development.

Even more rewarding is the realization that the Village's current plan for the redevelopment of one of the districts (the Harrison Street Arts District) in Oak Park developed by the Lakota Group,[3] a planning consulting firm, incorporates many of the key design and planning recommendations[4] made by the UIC Oak Park project. That the UIC team was able to help visualize the design for the district that eventually incorporated into an implementation plan is additional confirmation that the Planning Together process was credible.

The NIPC Common Ground process also resulted in many positive short and long term outcomes that further the participatory planning agenda. The short term success is evidenced by the awards and accolades the plan has received from the American Planning Association[5] and the Public Involvement Committee of the Transportation Research Board. However, NIPC (the land planning agency) merged with CATS (the transportation planning group) soon after the Common Ground process concluded. The success of the Common Ground work is that the commitment to participatory planning survived the agency merger and resulting organizational and staffing changes. CMAP has published their "Public Participation Plan" as required by federal regulations governing Metropolitan Planning Organizations. CMAP guidelines, developed for northeastern Illinois state:

- The public should have input in decisions about actions that affect their lives.
- Public participation includes the promise that the public's contribution will be considered in the decision-making process.
- The public participation process communicates the interests and considers the needs of all participants.
- The public participation process seeks out and facilitates involvement of those potentially affected by local and regional plans.
- The public participation process provides participants with the information they need to participate in a meaningful way.

- The public participation process communicates to participants how their input influenced the decision (CMAP, Public Participation Plan, 2007).

By emphasizing transparency, open communication, and accountability, these guidelines, as binding policy, go a long way towards establishing participation within regional planning processes. To a great extent, the success of the Common Ground process has allowed CMAP to be more innovative in preparing their participation plans for the upcoming forecasting challenges.

Summarizing from the three case studies discussed in Chaps. 5, 6, and 7 and by using the meta-evaluation framework that focuses on process design, short and long term outcomes, we can conclude that the introduction of participatory GIS activities gradually foster a more transparent and proactive planning process/practice, the closer they move towards the goals of an ideal participatory planning endeavor as previously described.

Notes

1. "Community Empowerment" is a slippery term. I further expand this discussion in Chap. 9.
2. Albrecht & Ramasubramanian (2004).
3. Lakota Group designs for Harrison Street available at: http://www.oak-park.us/Community_Services/Harrison_Design_Plan.htm
4. UIC plan designs for Harrison Street available at: http://www.oak-park.us/public/pdfs/UIC/10.31.03%20planningtogether3.pdf
5. American Planning Association: National Plan of the Year Award, 2006.

Part III
The Future of PPGIS

Chapter 9
PPGIS as Critical Reflective Practice

9.1 Introduction

The ideas in the book revolve around three cross-cutting themes – planning, participation, and technologies. What follows is a synthesis of lessons learned from the literature (Chaps. 2 and 3), the survey of the state of the practice (Chap. 4), and the individual case studies (Chaps. 5, 6, and 7) and an assessment of their relative merits (Chap. 8). In this chapter, I propose that GIS facilitates individual and community capacity building. Yet, the advantages that GIS offer are only maximized when the tools are embedded and integrated into a participatory process. The participatory process helps to empower individuals and groups because it allows for a balance between reflection and action – an approach I call critical reflective practice.

9.2 Planning Is Consensus Building

Although this book is focused on planning practice in the United States, it is important to remember that American neighborhoods and cities cannot remain isolated enclaves. Dramatic socio-demographic changes, waves of in-migration of skilled and unskilled workers, liberalization of cultural and religious attitudes, immigration, and even fear of terrorism and global pandemics have created complex physical and virtual networks that transcend conventional jurisdictional boundaries. In this context, accepted ideas about "good" planning practice are rapidly changing.

The rational-comprehensive model of planning was an expert-driven model (Alexander, 1992). Basically, this approach, one that defined much of post-World War II planning in the United States emphasized comprehensive strategies that emphasized the development and management of land (the physical environment). Planning in the 1960s and 1970s was almost Janus-like; on the one hand, the field emphasized sweeping changes to the status quo taking on the thorny issue of racial discrimination in housing, education, and employment. On the other hand, the field largely eschewed bold and dramatic physical design interventions, advocating incremental changes in an attempt to undo the negative consequences created by the brash and bold approaches of an earlier era.

L. Ramasubramanian, *Geographic Information Science and Public Participation*, Advances in Geographic Information Science, DOI 10.1007/978-3-540-75401-5_9,

In the 1980s, the profession floundered as it coped with the twin impacts of deindustrialization and deregulation. During this time, planning became emblematic of the problems created by government regulation that were seen as stifling the creativity and productivity of the private sector. During this time, large planning projects were developed and maintained by public-private partnerships,[1] essentially new institutions that were designed to promote entrepreneurship and development.

Since the 1990s, planners and planning institutions have stated that they attempt to build consensus, balancing the interest of the State, market, and civil society (Forester, 1989; Hoch, 1994; Healy, 1996; Innes, 1996). This ideal manifests differently, depending on the institutional actors, the scale at which consensus building activities occur, and the range of ideas and ideologies represented by the stakeholders. The planning-as-consensus building-model emphasizes deliberation and communication as well as a commitment to include the views of a wide range of individuals and groups, especially those most vulnerable and directly affected by any proposed actions. In this model, different interest groups are encouraged to reconsider their own positions on specific issues in light of the needs of other participants', avoiding stand-offs between competing positions.

The planning-as-consensus building model is an academic's dream, and a practitioner's nightmare because of the demands it places on planners and planning institutions.

9.3 Participation in Consensus Building Efforts

Firstly, consensus building requires the early and active engagement of diverse interest groups and a customized citizen participation process to meet a particular set of goals and objectives. Secondly, consensus building requires planners to be inclusive, specifically working to build trust and legitimacy for the process and the actors involved in the process. Finally, the consensus building ideal requires that different stakeholders work through their differences, actually solving problems along the way, rather than glossing over them.

The typical planning practitioner, particularly the middle-level manager charged with ensuring full and active public participation seldom questions the need to engage the public; the question on her mind is not *should I?* Rather it is *how do I*? Pondering this question creates incredible anxiety among professional planners given the responsibility of designing and implementing effective participation strategies. Throughout the book, I have provided strategies to cope with this anxiety – reminding readers that the design of participatory and consensus building processes deserves a lot more attention that it receives at present. One of the best strategies for success is to focus on creating a highly customized participatory process following the guidelines laid out in Chap. 3, rather than using a formulaic, paint-by-numbers approach to the management of public involvement. Participation in a consensus building process is not business as usual, it is a "qualitatively different process, one that puts citizens, business owners, environmental activists, builders, and developers, and many others in a collaborative venture with planners, planning commissions and the traditional power structure" (Klein, 2000, p. 438)

Consensus building is an information hungry endeavor. In most situations, people can agree on broad principles, for instance about the need to create new jobs or to preserve the environment. However, in envisioning the future of their neighborhood or community, two people who care about creating new jobs may have divergent views about how this goal should be achieved. For instance, one individual or group may argue that the town work to lure a single employer and provide them with big benefits to locate in the community, while another individual or group may advocate a different strategy, such as the creation small-business training program and the provision of micro-loans to start new businesses. In the end, both approaches may be necessary to fulfill the needs of the community, but each strategy requires the assemblage and analysis of different types of data and projections, and an analysis of resulting impacts. These facts and figures are likely to play a significant role in a successful consensus building process "the planners' task is to ensure that the information is credible, reliable, and understandable" (Klein, 2000, p. 434).

The twin trends – advances in information and communication technologies (GIS and the like) and the planning profession's adoption of consensus building model have co-evolved over the last two decades. They now address more effectively the dilemmas that the field has encountered in coming to terms with citizen participation, discussed previously in Sect. 1.3.

9.4 Framing Planning Issues

In Chap. 1, I observed that the way a problem is framed has a big impact on the solutions that are proposed. Using a metaphor of a story,[2] I argue, that GIS and associated digital technologies have made it possible for individuals, community groups, special interest groups, and planning agencies to tell their own story, from their point of view, and do so in a timely and cost-effective way. Each story that is told offers a different view of reality and "represents a special way of seeing" (Schön & Rein, 1994, p. 26). Each story selects and identifies different features and relationships that become the "things" in the story that are woven together to create a compelling tale.

In Chaps. 1 and 5, I observed that many urban renewal agendas (regardless of the underlying political or racial motivations that may have fueled them) are publicly framed by professional planners as problems associated with the built environment. Mel King reacts by saying, "Labeling those streets as slums depersonalized the issue, and blocked out any understanding of the impact urban renewal would have [on peoples' lives]" (King, 1981, p. 21).

Not knowing what the official planners in the 1960s actually did, I can only speculate that they used data, maps, and analyses about age of housing stock, the number of residents who lived in each dwelling unit and the ratio of open spaces to built-up area in the South End to make the case for overcrowding, concluding that there were far too many residents living in each dwelling unit than was acceptable by "official" standards of health and hygiene, and that there was a shortage of parks and other open spaces. They are likely to have had "data" about each of the features they selected to use and a method for establishing relationships between selected features.

Community groups in the 1960s may not have had access to the same data about housing stock or housing quality as the city officials, except in situations where a radical planner took the data out of the planning offices and "leaked" it to community groups who were mobilizing against official actions. What community groups did have on hand were oral histories, photographs, home movies, and memories of people and places in the neighborhood that were special. These human interest stories may have been used occasionally to solidify support among members of the public already sympathetic to their concerns. However, the main strategies to challenge these official acts of tyranny were through organized protest and resistance.

Can 1960s style urban renewal happen today? Unfortunately, the answer is *yes, it can*; in fact, recent evidence suggests that these approaches to planning and development occur quite regularly in large and small cities. In 2003, a private developer, in partnership with the city and State of New York proposed to develop a mega complex of offices, luxury housing, and a new basketball arena in a section of downtown Brooklyn above rail yards owned by the Metropolitan Transportation Authority. The proponents of the project put forward many reasons to support their super-sized development agenda. One reason that was repeated often was that the neighborhood was "blighted", observing that there were many vacant properties in the neighborhood, that the area was under-utilized and thus in need of revitalization. It was not quite apparent until I started doing some research about this issue that one of the reasons for the high vacancy rate was that the developer had systematically and quite legally acquired much of the property within the proposed development footprint. The argument about blight and the subsequent invocation of eminent domain authority to condemn the area as blighted and in need of revitalization appears to have been a strategy to counter the resistance of those tenants and owners who had refused to sell and make way gracefully for the new development.

As GIS and other digital technologies have become more accessible, citizens have learned to use these tools to further their own cause célèbre. In the Atlantic Yards case for example, digital technologies were used in many ways to reframe the issues – a documentary about the resistance called Brooklyn Matters[3] focuses on how race and class are used as wedge issues to divide and dilute community opposition to the project. My colleagues and I at the Hunter College Center for Community Planning and Development used GIS and census data analyses to counter the preposterous claims of project developers that the massive developments (4,500 new units of housing) they proposed would not put pressure on city services such as emergency services and public spaces in the neighborhood, and schools. Hunter College planning students monitored traffic flow in the neighborhood (the site is situated at the intersection of three major thoroughfares) to estimate how the new developments, one of the densest development proposals in all of New York City would affect traffic conditions. Other community groups created three-dimensional renderings to show how the massing of the buildings, when built, would leave many parts of the neighborhood in permanent shadow. A website sponsored by the main opponents of the project, a coalition called Develop, Don't Destroy Brooklyn (DDDB),[4] became quite adept at sending out information and updates about planning meetings, protest actions, legislative strategies – in short serving as the virtual meeting place for the opposition.

The opposition to the Atlantic Yards project showcased the comparative advantage of using a wide range of digital tools (GIS, PhotoShop renderings, 3-D visualization, and video to mention a few) to make it possible for individuals and groups to frame and re-frame planning problems in order to share an alternative vision of the future. In particular, proponents of alternative community-based plans such as the Unity Plan developed by the community, under the leadership of an architect, and with the support of a local city council member, were able to share their ideas with the general citizenry and expand their reach using the Internet. The drawings, maps, and analyses can be accessed through the DDDB website and are available to anyone who wants to review them.

In general when multiple framings of an issue are aired, elected officials, professionals, and ordinary citizens have an opportunity to reflect, perhaps reconsider, or clarify their own positions about key issues. In the case of the Atlantic Yards project, the controversies were eventually reported in major newspapers including the New York Times. Many non-Brooklynites gradually became aware that there was not complete consensus within the community about the Atlantic Yards project. As of writing these pages in 2009, the Atlantic Yards project is stalled – in part because of the steady onslaught of community opposition, and in part, because of the 2009 economic downturn. I argue that the use of digital technologies and media in challenging the Atlantic Yards helped community organizers achieve their goal to slow the juggernaut; what once looked like a "done deal" is no longer looking as secure.

Within a participatory framework, GIS and other digital technologies facilitate and allow for the expression of multiple narratives that can flow and intersect, addressing concerns related to efficiency, economic growth, aesthetics, historical heritage, sense of place, culture, morality, ethics, and the environment. Within the planning-as-consensus building framework, these narratives can be heard without privileging one view over another, a first step to forging agreement about particular decisions regarding the future. However, because so many narratives are possible, facilitators of consensus-based approaches to planning have an additional obligation to assist citizens in sorting through multiple narratives by highlighting points of convergence and divergence.

9.5 Advocacy for the Public Interest

One of the dilemmas that affects participation in consensus building activities concerns representation – the need to bring a diverse group of stakeholders into the planning process. Many times, professional planners are ridiculed for serving the interests of the powerful, whether they are loudest protesters (the special interest groups) or those individuals who have friends in high places.

Academics writing about PPGIS remind organizers of their obligation to assemble a diverse group of participants into any planning process and to do it as early as possible. For example, Schlossberg and Shuford (2005) propose that decision makers, implementers, affected individuals, interested observers, and members of the public at large (random public) should be consulted in any process, although the obligations of engagement may vary, depending on how removed they are from the project/problem at hand.

Digital tools now make it possible for citizens from different walks of life, including those who are usually relegated to society's margins to emerge and engage within the public sphere – consider the recent phenomenon, where a homeless man in Houston, Texas, can tell his story on the Internet[5] and raise money and social support as he struggles to rejoin mainstream society. John's story and its associated media coverage may force the hand of elected officials, professional planners, and community activists by drawing attention to the larger issues of homelessness in their city and community.

PPGIS advocates define the term "public interest" by asking: who is the public? Who should participate? I propose that we are better off asking a different set of questions: Who should the professional planner/the community activist/grassroots group advocate for? How best should they do it? What tools and skills do members of the public need to help them articulate their advocacy positions? At this juncture, it is useful to remember that the needs of future generations (children and youth) also require strong advocacy. Typically, professional planners are supposed to represent the interests of those who are absent or rendered voiceless in some way. This obligation should not disappear in a consensus planning model.

The planning-as-consensus building model argues that everyone, who has a stake, has to be engaged in the planning process. That is the theory. In practice, by adopting the most instrumental goals of participation – as leading to open and honest exchange of ideas, this model can make it easy for planning agencies to neatly sidestep the issue of power. In other words, it allows planners and agencies to avoid any consideration of the structural imbalances that are built into formal planning frameworks, even those that embrace a high level of citizen involvement. I am very firm in my belief that merely bringing people together and asking them what they want is not planning, it's a feel-good exercise at best, and disrespectful and offensive at worst.

Planning as consensus building requires the identification of shared interests, the sharing of credible information, skills, and leadership and should not waste people's time (Klein, 2000). To illustrate, how understanding public interest in this way works, let us return to the case of the South End Housing and Planning Coalition, the subject of our case study in Chap. 5.

If you remember, when the City of Boston and private developers proposed SETSA, a bio-medical complex in the South End (discussed in Chap. 5), they used an economic argument to locate the development in the South End. The developers argued that biotechnology was a growth industry with a potential to generate many jobs. They proposed that low-skill jobs and economic gains from smaller spin-off companies would trickle down to South End residents. SEPHC was not convinced. They had more questions than answers, and felt that they needed to undertake their own research before finalizing their policy positions. SEHPC's research eventually reframed the problem as a mismatch between the skills of poor South End residents and the employment opportunities being offered by the new project. SEHPC argued that the new jobs generated through the project would benefit only highly skilled professionals and adversely affect the employment opportunities of South End residents with limited resources.

The SEPHC research used data from various sources to answer the specific research questions that concerned the coalition. Data from the U.S. Census, research reports and statistics compiled by the Boston Redevelopment Authority, unpublished theses and dissertations and various newspaper articles and books were used to develop a coherent understanding of Boston's economy and industry profile on the one hand, and the development of biotechnology firms on the other hand. This is not to say that the developers and/or the city did not use the same data. However, SEPHC used the data to answer different questions, for example, how many low and semi-skilled jobs would be available to South End residents? In addition to data analysis, interviews with representatives of biotechnology firms and community advocates assisted the SEPHC in developing a comprehensive picture. For example, they discovered that although in general, it was accepted that biotechnology development would spin off projects that required low and semi-skilled workers, the prevalence of hundreds of colleges and universities in the Boston area combined with the high level of unemployment at that time would have resulted in college graduates applying for and receiving jobs that would have been filled by high school students with appropriate training in other cities.

Thus, SEPHC's knowledge about the biotechnology industry, the number and the type of jobs that it could create, the skill level required for such jobs in Boston and the limited availability of training and educational programs for potential job-seekers in the area's community colleges and schools allowed the coalition to negotiate for other benefits, rather than focusing on the creation of jobs for South End residents.

Some might conclude that SEPHC had "accommodated" the interests of the City by moving away from its original claims. I argue that the SEPHC moved away from an oppositional stance towards building consensus, because the SEPHC leadership had the best interests of the community, particularly its most vulnerable residents in mind. SEHPC's report was path breaking in that it was the first time that a community group had framed a problem differently from the city, shifting the frame of reference slightly and supporting their arguments with data and information. By doing so, they were also more effectively able to advocate for the different types of stakeholders (publics) in their community. In the preface to the final report, SEHPC's Executive Director wrote:

> When SEPHC first became involved with the development of the biomedical project . . . the community process attendant with that development did not yield enough specific information regarding employment opportunities for us, and the community coalition we work with, to engage in a meaningful dialog with the developers regarding job opportunities. Individual residents or entire neighborhoods, in dire need of employment opportunities must not become victim of false expectations . . . We have been asked if this is not an unusual endeavor for SEHPC. It should not be. This isn't the '60s, 70's or even the 80's. . . . Community groups must become more self-reliant, developing techniques to match the times (SEHPC, 1991).

Let me offer another example to explain how a planner can be very effective when advocating for the public interest. Most public agencies will acknowledge that the simplest strategy they can use in providing access to information is one-way information provision, a very rudimentary form of outreach. Even Sherry Arnstein (1969), who dismissed these types of outreach efforts as tokenism has acknowledged that there is always a need for such information dissemination – we cannot get around the reality that in order for people to become involved, they have to know what's going on.

Once an agency has made a commitment to disseminate information, the next question that needs to be addressed is – *how do you get the word out?* In the pre-Internet world, the answer would have been *the daily newspaper*. For the last half century, the most cost-effective way of informing the general public about a meeting or community event would have been to place advertisements in major newspapers. Nowadays, public participation and outreach specialists observe that the newspaper is one of the least effective ways of doing community outreach. Given the decline in daily newspaper circulation, their conclusions are not surprising.

At the same time, community newspapers, particularly non-English language newspapers that cater to the needs of immigrant communities continue to be popular and attract a wide readership. Again, this is not surprising; the people who read these (non-English) newspapers may not have the luxury of a desk job with unrestricted access to the Internet, or the flexible schedule that allows them to get the day's news on-the-go. For these people, the newspaper is an artifact that is read on the train, while waiting for the bus, shared among friends and family members, and often a way to stay connected simultaneously to local (what's going in my neighborhood) and global news (what's going on back home in my village, whether that village is in Mexico, Dominica, or India).

If community advocates are truly concerned about the public interest, then they are unlikely to take an either-or strategy. They recognize and use the innovative contributions of digital technologies,[6] but are also aware of the digital divide that mirrors more persistent divides of social class and education. They will work to establish connections with those individuals or groups who do not have access to digital technologies. They will always attempt to educate and advocate for those community cohorts, who are not familiar with the history and context of how planning occurs.

9.6 Managing Participation

By many accounts, current approaches to managing participation are not very effective (e.g., Day, 1997; Innes & Booher, 2004; Ozawa, 1991; Yankelovich, 1991). This is particularly true, when public participation is a formal requirement. Established methods[7] of soliciting and managing participation are constantly decried for causing frustration, anger, and mistrust.

An undercurrent that runs through some of this literature, particularly in the PPGIS literature is a belief that the participatory planning agenda is doomed because narrow and deep interests represented by organized groups will always dominate, overshadowing the broad and shallow interests of ordinary citizens. I dismiss this concern as an academic preoccupation, one that is potentially divisive because it seems to reject the value of community organizing. In other words, it would appear that we, as a society, are intrigued by the idea of the accidental activist,[8] one who rises up to protest injustice and transforms a staid/oppressive situation through their intrinsic leadership skills and erudition. However, the most successful accidental activists have been inspired by, or have learned from the experiences of organized groups and campaigns. In other words, our goal in managing participation should be focused on constructive engagement with individuals and groups in ways that make it possible for participants to take ownership of the process.

King (1981) and Susskind (1983) have argued that in most communities, citizens adapt and learn how to participate effectively in planning processes over a period of time. Learning within this process occurs initially because of access to information and the literacy necessary to comprehend and process that information. Subsequently, social interactions and community engagement help to transform that information into intelligence/knowledge that can be used in the context of collaborative problem solving. Thus, most individuals and groups tend to move from a "service stage", defined by paternalism, through an "organizing stage", which creates many tensions and conflicts, eventually to "building institutions" that can effectively represent local community perspectives in a respectful an trustworthy way. At this stage, individuals and communities are more willing to work collaboratively with governmental entities because they have positioned themselves as co-equal partners engaging in dialogues for their mutual benefit.

Following this reasoning, I maintain that the management of participation within a consensus-building model must include methods and strategies of capacity building. We will discuss capacity building in some detail in Section 9.8. For now, we can conclude that participation is essential for decision makers to identify public preferences, to hear the voices of marginalized groups and to create legitimacy for the decisions that are ultimately made. For community groups, participation can create new networks of association and personal relationships, build trust in government, and strengthen the group's institutional capacity.

9.7 How Technologies Have Evolved

Advances in information and communications technologies like GIS were originally promoted as a way to create new efficiencies, to speed up routine tasks and analyses and to allow us to make better use of our time. As the information needs associated with planning have grown, planning practitioners have explored the benefits of using digital tools to transform and energize conventional participatory activities. GIS, in particular offers many benefits to facilitate communication among different groups. These include the ability to:

- *Identify and clarify spatial relationships*, e.g., what is the spatial proximity of the new development to the stand of old growth forest that we are trying to preserve?
- *Speed up information processing time to answer formal criteria-based queries in real time, or close to real time as possible,* e.g., to answer questions such as, how many land parcels in the community have the same land use designation, and thus are likely to be affected by the policy change that is currently being proposed?
- *Improve communication with and among non-specialists*, e.g., how have settlement patterns changed over a 5, 10 or 20 year time period and how have these patterns affected businesses in the community?
- *Create what-if scenarios, e.g.,* what exactly would happen to tax revenues, our water supply, air quality, traffic conditions if we increased the density in this neighborhood by 100%, 50%, or 10%?

Adaptations to conventional GIS systems now allow users to experience these relationships through 3-D renderings and immersive simulations that place the

Fig. 9.1 CommunityViz rendering comparing different scenarios

user in the middle of a scene that they can navigate by walking, driving or flying through. These enhancements make it possible for end users to appreciate qualitative attributes of the environment, such as the impact of light, shadows, noise, and pollution, all elements that are hard to describe but relatively easy to understand experientially (Fig. 9.1).

Other digital technologies such as internet-based group communication tools and electronic keypad polling also help to make the ideals of public deliberation, debate, convivial conversations, arguments, and collective problem solving work in the real world (Fig. 9.2).

Nowadays, GIS and digital tools facilitate and improve communication in collaborative settings because they can link and access different types of information in ways that are most meaningful to the end user. Yet, most readers will intuitively recognize that having access to the tools and the data will not help citizens and vulnerable groups make informed choices. In order to make reasonable decisions, citizens (stakeholders) still need to understand the relationships between choices and consequences, for instance, what would happen if these plans are not implemented? Who would be adversely affected?

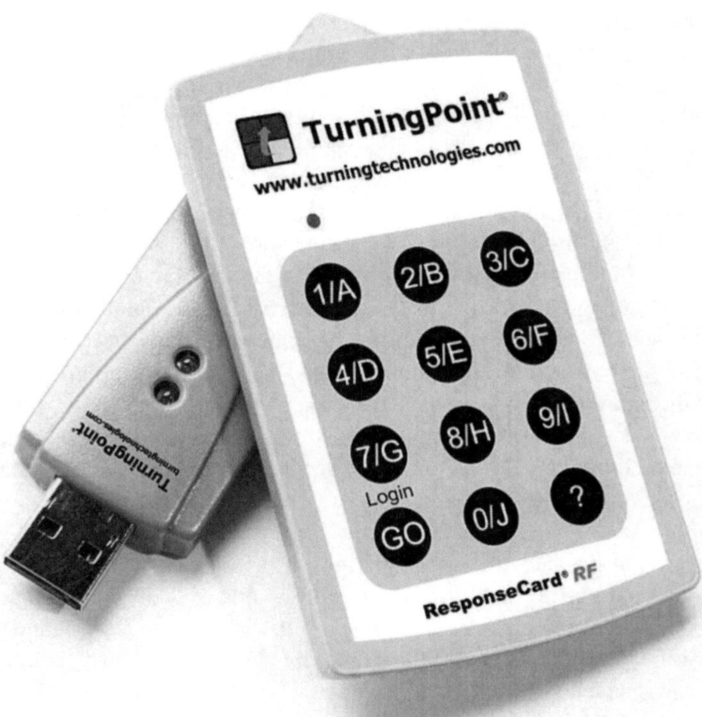

Fig. 9.2 Keypad polling

In Chap. 3, I noted that an ideal PPGIS activity should build capacity of participants by providing them with access to information and the support needed to make sense of it, fully engage participants in all aspects of the planning endeavor, develop techniques to integrate participants' views into the planning process, and provide transparent strategies for data dissemination. Based on the discussion we've had thus far, an ideal participatory GIS activity can be elegantly embedded into the planning-as-consensus building model. Now, everything seems to hinge on the goals of capacity building. How do you build capacity among citizens who do not have formal education? Can they understand formulas and models? Can they be expected to provide feedback about arcane planning decisions?

9.8 Building Capacity and Empowerment

Based on my experience over the last 20 years, I've learned that some individuals and groups are able to creatively use digital technologies to make these thoughtful judgments. When citizens are involved in participatory processes over a period of time, they do learn the specialized vocabulary and jargon that planners and other analysts use, they learn how to read the zoning code, decode demographic and map data and so on. They come into the process with quantitative, digital, and spatial skills and have no difficulties using them within a collaborative problem solving process. Prolonged engagement in community affairs brings its own rewards and responsibilities. The rewards include gaining a better understanding of formal planning activities, being treated with respect by official planners, and, expanding personal networks of association. Eventually these gains can translate into a feeling of accomplishment and empowerment. Such empowered individuals eventually become more active in their communities. Groups engaged in these kinds of community planning activities with the intensity of involvement showcased in the case studies learn to put GIS and other technologies in their place, in other words, they can use the power of GIS and other digital tools while being cognizant of the problems that may result – e.g., that the organization may become very focused on data driven decision making which will limit its effectiveness when there is no data available. In short, they are able to use the technologies in liberating ways, while minimizing its potential to create repressive outcomes – they are empowered users!

Empowerment has become common argot in the past decade, replacing a more easily understood but politically charged word – power. Robin Denise Johnson's (1994) book, *Where's the Power in Empowerment*? provides a working definition of empowerment to initiate our exploration of the theme. She defines empowerment in an organizational setting as the sum of authority, control, and trust. Specifically, "Empowerment is decision making authority, appropriate internal control of tasks and time, and responsibility for results of decisions and actions due to organizational and managerial support that engenders interpersonal and intergroup trust." (Johnson, 1994, p. 11).

Using Johnson's definition of empowerment as a reference point, I declare that it is almost impossible for an outside entity to bestow decision making authority to an individual or group because that very gesture will become disempowering. In reality, that individual or organization is likely to acquire decision-making authority by developing the confidence, skills, and experience to make decisions and explore the limits and boundaries of those decisions in an environment that supports organizational learning. Thus, we may conclude that while many individuals, government and non-government organizations seek to empower another person or group, their interventions can, at best only, influence the psychological, social or political contextual conditions likely to affect the other person or group. It is possible that changes in contextual conditions have little or no impact on the individual or group that is the target of empowerment initiatives.

John Friedmann (1992) conceptualizes empowerment as an interconnected triad of social, psychological, and political power. Social power is concerned with an individual's or group's access to certain resources of production, such as information, knowledge, and skills. Political power is the access to decision making processes, particularly those that affect an individual's or group's own future. Psychological power is an individual sense of potency, demonstrated in self confident behavior. It is often the result of successful action in the social and political domains.

Friedmann's conceptualization of empowerment supports Mel King's (1981) conceptualization of different dimensions of power, particularly its effects on communities of color and communities with limited resources. King argues that the *structure* (social power) is imposed by outside forces who tend to control available resources, while the *image* (psychological power), although internal to the community, is partially influenced by external forces like the media. King proposes that redistributing or *balancing* power (political power) will determine how decisions are made inside and outside the community.

Elizabeth Rocha (1997) developed a ladder of empowerment, creating a typology to organize and disentangle empowerment theory literature. In the same way that the Arnstein ladder allowed us to explore variations in citizen participation, Rocha's ladder creates steps that move from individual to community empowerment, where each step in the ladder indicates a progressive advancement in the pursuit of empowerment. Rocha moves from atomistic individual empowerment embedded individual empowerment, mediated empowerment, socio-political empowerment, and political empowerment. Friedman and Rocha appear to come at the issue of empowerment from opposite ends; for Friedman, psychological empowerment is an outcome of working towards social and political empowerment, while Rocha argues that psychological empowerment is a necessary first step in the pursuit of political empowerment.

Sarah Elwood (2002) proposes that we unpackage empowerment to consider its distributive dimension (the achievement of tangible material changes/outcomes), procedural dimension (the transformations of processes in order to legitimize the contributions and/or needs of citizens' groups), and its capacity building dimension (the ability of citizens to take action on their own behalf).

I propose that much of the confusion about what empowerment means is associated with empowerment efforts that are initiated from the outside. External change agents believe that the individual or group being "empowered" will gain some control over their immediate economic and/ or social situation due to presence of favorable external conditions. While outside initiators of empowerment initiatives agree that there are great benefits to be gained in creating empowering conditions, the significance they attach to these conditions differs widely depending on their political world views, their professional backgrounds, and personal experience. For example, Peterman (1996) identifies at least three meanings of empowerment that emerge when activists and policy makers support the idea of resident/tenant managed, publicly funded, and subsidized housing. Each meaning supports the idea of resident or tenant management for different reasons. Conservatives advocate home ownership as a means of fostering empowerment. For them, resident management of public housing is a step towards home ownership – the final goal (the distributive dimension). The focus of empowerment efforts is the individual. At the same time, the liberal position advocates shared decision making and dual management strategies in public housing. Within this position, the idea of publicly funded housing is not rejected and home ownership is not seen as the final goal for all residents. Instead, liberals concentrate on creating inclusive and collaborative processes of governance, which are expected to create empowered citizens or citizens' groups (the procedural dimension). The progressive meaning of empowerment shifts the debate from individual home ownership or rentership and equates empowerment with community organization and community control (the capacity building dimension). Viewed from this framework, the development of strong community-based leadership is seen as an outcome of empowerment initiatives. Resident management is considered only one of many possible solutions likely to emerge when a community is collectively engaged in determining its future.

To conclude the discussion about empowerment, I propose that it is imperative to link psychological (individual), social (group), and political (institutional) power to any definition of empowerment. *Empowerment is the process and outcome of critical reflective practice*. Empowering instances, moments, activities, and/or outcomes emerge when individuals or community groups negotiate the interconnected triad of psychological, social and political power. In order to do this, they navigate the dialectic between action (activism) and reflection (research and analysis). Critical reflective practice consciously links action and reflection – praxis (Freire, 1970) without ignoring structural disparities (Friedmann, 1987, 1992). In order to become an empowered user, an individual or group must be willing to invest time thinking about the issues as well take concrete steps to address specific concerns. Doing one or the other is not an option.

Empowerment is not a fixed resource, rather it is a process that is constantly evolving and changing. For example, an individual who is able to examine her actions and reflect upon them, while taking into account the larger social and political context in one instant or situation, may be unable or unwilling to engage in the same process at another time for a variety of reasons. Therefore, the same individual or organization can act in an empowered manner in one situation while appearing disempowered in another. Empowerment is not viewed as an end in itself to be attained by the creation of certain favorable conditions but rather as a guiding principle that forms the basis of problem solving and decision making throughout the life of an individual or organization.

The long term goals of participatory planning or consensus building endeavors are to foster/improve the social, intellectual and political capacity of participants and other affiliates; to create ways for participants to become articulate and effective advocates for both individual and collective interests; to better prepare citizens to participate fully in urban governance, and, engender trust in participatory processes and their outcomes.

I have placed high expectations and demands on those planners who are charged with fostering public engagement within consensus building frameworks. Committed practitioners will discover that when they uphold the ideals of critical reflective practice, they are more effective advocates themselves. While reflective practice implies that individuals or organizations act and then reflect on their actions, incorporating an analysis of power transforms the entire reflective process. Over time, planners will be able to develop a set of principles to increase the emergence of newer empowering moments or situations. As planners become more effective advocates for their constituents, they will be recognized as community leaders.

Notes

1. The Central Park Conservancy, http://www.centralparknyc.org. According to the Conservancy's website, the Central Park Conservancy is a private, not-for-profit organization founded in 1980 that manages Central Park under a contract with the New York City Department of Parks & Recreation. The Conservancy provides 85% of Central Park's $27 million annual operating budget and is responsible for all basic care of the Park. The Battery Park City Authority, http://www.batteryparkcity.org, According to the Authority's website, the Hugh L. Carey Battery Park City Authority is a New York State public benefit corporation whose mission is to plan, create, co-ordinate and maintain a balanced community of commercial, residential, retail, and park space within its designated 92-acre site on the lower west side of Manhattan. The Authority website states that public-private partnerships will continue to be the model for private sector development utilizing a competitive public bid process to optimize value.
2. Also see Throgmorton (2003).
3. Brooklyn Matters, http://www.brooklynmatters.com
4. Develop, Don't Destroy Brooklyn, http://www.developdontdestroy.org
5. See http://pimpthisbum.com
6. Some planning agencies use non-traditional methods such as creating groups on FaceBook to engage citizens who are not usually involved in participatory planning activities
7. Established methods include public hearings, plan or design review, submission of verbal or written comments, participation on advisory committees or task forces.
8. Rosa Parks was active in the Civil Rights movement, although she is often portrayed as an accidental activist, Mahatma Gandhi was a highly educated lawyer; his decision to pursue Satyagraha (nonviolent protest) was strategic, not whimsical.

Chapter 10
Where to, from Here?

To really listen means there's a willingness to change!
Comment at a community meeting

10.1 Introduction

Planners and planning activities are inherently political; consequently, we must understand our work as individuals, as part of groups, and as part of the institutions. It shapes how we view planning issues, the methods we identify as appropriate to analyze and solve the problem, and the outcomes we consider desirable or favorable. In other words, our individual agency and motivations are mediated and influenced by the group (societal) settings within which we work, and are ultimately shaped by the larger institutional and political constructs that govern our lives. It is in this context that we must understand the dynamics of doing participatory planning enabled by digital technologies.

The goal of all planning is to create options and choices to solve the problems or issues that we are currently experiencing. Consensus building, the planning approach, described in Chap. 9 provides us the rationale and the approach for working together. Essentially, consensus building relies on logic and reason, the power of deliberation, and the power of civic engagement. Colloquially, consensus building believes that if enough of us spend time talking things through, cooler heads will prevail, and we can collectively figure out how to solve the problems that we are facing. This model can apply to the household, the neighborhood, the community, the city, and the region.

In my book, I propose that the consensus building model can benefit immensely from the power and promise of digital technologies such as GIS. However, I am also arguing that the tools by themselves are ineffectual and sometimes counterproductive. What is really needed is a good process that marries the power of the tools with the techniques and approaches that can operationalize consensus building and make it work in practical ways.

L. Ramasubramanian, *Geographic Information Science and Public Participation*, Advances in Geographic Information Science, DOI 10.1007/978-3-540-75401-5_10,

Digital tools have evolved and adapted themselves to serve societal needs in a somewhat haphazard fashion. It is time for planning practitioners to seize the comparative advantages afforded by digital tools and integrate them more effectively into consensus building activities. To achieve this integration, I have developed a set of guidelines for planning practitioners seeking to use digital technologies as a part of planning processes.

10.2 Guidelines for Participatory Planning with GIS

1. *Put people first, the technologies come later.* This is the simplest but most important organizing principle that you should hold on to, regardless of your particular role in the planning process. I sometimes hear students or practicing planners say, "I just want to remain behind the scenes" or "I want to work on the hard/tangible stuff – numbers, data, that sort of thing". Unfortunately, planning processes don't work this way, at least not the kind of planning processes we want to create and sustain. The choice of technologies is less significant than ensuring that participants are engaged in the planning process from its inception to implementation.
2. *Start planning with the people and tools you have.* No planning process is perfect out of the box. Planning practitioners are understandably very anxious about assembling the right mix of participants, aspiring for that perfect balance of diversity, commitment, and competence among participants. Yet, any planning issue naturally brings together a group of people who coalesce around the topic spontaneously. Begin work by engaging this group, and then gradually extend your circle of influence by inviting new members and creating appropriate participatory opportunities. To reject the people who showed up and expressed concern about an issue, because they are not the "right" kind of people, for whatever reason is churlish and disrespectful. As a corollary, start your meetings on time, honor the people who came on time and find a way to use their time meaningfully while you wait for the late arrivals.
   With regards to technologies, most planning processes can begin with a laptop that runs basic software like word-processing (Word), spreadsheets (Excel), and presentation software (PowerPoint). Consider using free-to-use tools to create stakeholder groups online. These online forums include additional free-to-use tools to help stakeholders collaborate on document-editing, participate in group "chat" sessions, and store electronic files in a common location. Those in need of maps and data can access much of this information online. Initial planning activities can use Google Maps that can be customized and used for collaborative planning. You can add new tools as the process evolves.
3. *Be creative yet cautious in developing partnerships.* Most participatory planning processes are under-resourced. Often, the budget may only allow for one community forum, when there may be a need to hold three and so it

will go. Your goal must be to first design the best possible process to serve your needs and then scale it and adjust it so that it fits within your budget. Most participatory planning activities rely on volunteer support to succeed. Typically partners include students, university professors and software developers. Philanthropic foundations, business groups, and governmental agencies may provide funding opportunities. Be very creative in reaching out to partners to assist you in achieving your goals. But, be cautious. You must be very clear on what your collaborators or partners want as a result of their involvement with the process. Are they asking for data that they can use in their research? Are they seeking academic credit for being involved? Are they using your project to test their new software? Each of the questions I've raised have ethical implications that can facilitate or hinder your efforts at building and sustaining a credible participatory planning process. I believe that partnerships are essential for success, but I also emphasize that the partnerships must be very explicit and transparent about the needs and expectations of all the partners. Remember that the citizens (the participants) are also a partner in these negotiations.

4. *Educate to empower.* You must build the capacity of your participants in order that they can be engaged and make useful contributions to the process. The discussions about context, purpose, and goals, "why is this so important for us to undertake at the present time", must precede the instrumental discussions about how can we achieve our goals "these are the programmatic steps that we need to take". The particular acquisition of skills and knowledge to undertake specific tasks is a necessary step, but it should come at the end. For instance, not all participants in a collaborative planning endeavor need to use mapping software, but they must all understand why the tools need to be used in particular ways to answer the questions they have raised. We have to get rid of the notion that computing is a black box. These educational activities, particularly the discussions of purpose, context, and goal setting are best accomplished through peer-to-peer knowledge sharing, rather than through conventional instructional models.
5. *Participation is about creative problem solving, not data processing.* Be respectful of participants – do not use them as free or cheap labor to collect data, attend meetings at times that are inconvenient for you, or transcribe meeting notes. Thus, in the case of data generation, you should engage participants in a discussion about what data must be collected and put in a good faith effort to assemble the data yourself. If you are not able to do the data collection, then ask yourself, what the problem is. If you are under-resourced, ask for volunteers but be explicit about why they are collecting the data and not you. While data gathering by citizens can be very effective in building community or to create peer-to-peer information sharing opportunities, I propose that the best use of participants' time is to engage them in articulating questions and identifying pathways to answer those questions.
6. *Be systematic in engaging hard-to-reach populations.* Remember to bring in people who are usually left out of planning activities. Children, elders,

non-English speakers, new immigrants represent some but not all of these constituents. It is important to be systematic in engaging these populations over the entire time frame of the planning process.

7. *Be transparent and accountable.* Transparency (not complete agreement) is one of the keys to maintaining your credibility and trust in the community. Individuals and groups who are involved in the planning process must understand why you made particular decisions, and who you consulted before making those decisions. It is impossible to secure complete agreement, but usually, being transparent and accountable (taking responsibility for the decisions) are essential ingredients to securing the trust of the community. Upholding the principles discussed in Item 3 (partnerships) will resolve many problems related to transparency and accountability.
8. *Be rigorous in your research and analysis.* Do the research, every time. Don't build your work on faulty assumptions or old data. Remember that official records are not always correct, nor are residents' narratives always reliable. Yet, we often defer to the official record, and are much less inclined to acknowledge the validity of experiential knowledge. Being rigorous in a participatory planning process lies in knowing how to assess the validity and reliability of *all* information that you can assemble, regardless of its particular source.
9. *Have a positive, hopeful attitude.* Enter any participatory planning process with the belief that everyday citizens can come together and make good decisions about the future of their community. Remember, that they are as vested as you are in the success of their neighborhood and community, perhaps even more than you are.
10. *Participation is not a free-for-all.* Participation, in order to be meaningful and successful must be managed carefully. The more attention that is paid to the logistics of creating a good participatory experience, the better the outcomes will be. Good facilitation, time management, availability of appropriate expertise, and information, and providing for the physical and psychological comfort of participants is central to achieving positive outcomes.
11. *Act glocally.* Problem solve at different scales – consider the needs of the household, the neighborhood, the community, and the region as you work through the planning process. This approach will foster coalition building and new alliances, tap into research, data, and evidence to support your arguments, and help address the long term goals of participatory planning endeavors – to empower people psychologically and politically so that they can create systemic change.
12. *Have patience and good humor.* Your success and the success of the process depends on your ability to be patient and handle the most difficult situations with kindness and good humor. As a veteran of many participatory planning processes, I can assure you that there will be times and days when things do not work well. It is at these times that your leadership, your trust in yourself and in the intrinsic value of the participatory process, and your adherence to the principles 1 through 11 are most essential and relevant. Patience and good

humor can help you get through the trying times, but the trying times inherently a necessary part of the process for you and the community. Participatory planning, with or without the use of digital technologies, is difficult, messy, and time-consuming work. But it is work that must be done to achieve the goals of fairness, equality, and social justice.

In conclusion, planning practitioners must remember that we are in this business for the long haul, and therefore be willing to participate in the ebb and flow of the participatory planning approach. The goal of consensus building and participatory planning is to create and sustain better futures for all of us, a goal that can only be achieved when we work from a position of love and caring about others, and recognizing our humanity as a tie that binds us together.

References

Agre, P. E., & Schuler, D., (Eds.). (1997). *Reinventing technology, rediscovering community: Critical explorations of computing as a social practice*. Greenwich, CT: Ablex Publishing.

Akalimat, A., & Gills, D. (1989). *Harold Washington and the crisis of Black Power in Chicago*. Chicago: Twenty-first Century Books and Publications.

Albrecht, J., & Ramasubramanian, L. (2004). The moving target: Learning from GIS-based surveillance. *Journal of Medical Systems, 28*(4), 369–382.

Albrechts, L., Healey, P., & Kunzmann, K. R. (2003). Strategic spatial planning and regional governance in Europe. *Journal of the American Planning Association, 69*(2), 113–129.

Alexander, E. R. (1992). *Approaches to planning: Introducing current planning theories, concepts and issues*. Philadelphia: Gordon and Breach Science Publishers.

Alinsky, S. (1971). *Rules for radicals: A practical primer for realistic radicals*. New York: Random House.

Al-Kodmany, K. (2000). Extending GIS to meet neighborhood planning needs: Recent developments in the work of the University of Illinois-Chicago. *URISA Journal, 12*(3). Retrieved from www.urisa.org

Angotti, A. (2007, February). Plan NYC 2030. *The Gotham Gazette*. Retrieved from http://www.gothamgazette.com/article/landuse/20070206/12/2095

Angotti, T. (2008). *New York for Sale: Community planning confronts global real estate*. Cambridge, MA: MIT Press.

Argyris, C., Putnam, R., & McLain Smith, D. (1985). *Action science: Concepts, methods, and skills for research and intervention*. San Francisco: Jossey-Bass.

Argyris, C., & Schön, D. (1991). PAR and action science compared. In W. Whyte (Ed.), *Participatory Action Research* (pp. 85–96). Newbury Park, CA: Sage.

Arnstein, S. (1969). A ladder of citizen participation. *Journal of the American Institute of Planners, 35*(4), 216–224.

Audirac, I. (2005). Information technology and urban form: Challenges to smart growth. *International Regional Science Review, 28*(2), 119–145.

Barber, B. (2004). *Strong democracy: Participatory politics for a new age*. Berkeley: University of California Press.

Barkey, J. (2006). *Atlantic yards photo simulations*. Retrieved from http://www.pbase.com/atlanticyards/simulation

Barndt, M. (2002). A model for evaluating public participation GIS. In W. Craig, T. Harris, & D. Weiner (Eds.), *Community Participation and Geographic Information Systems* (pp. 346–356). New York: Taylor & Francis.

Beamish, A. (1995). *Communities on-line: Community-based computer networks*. Master in city planning thesis, Department of Urban Studies and Planning, Massachusetts Institute of Technology. Retrieved from MIT Libraries.

Bentrup, C. (2002a, July 3). REDCOOP pushes for advisory referendum. *Oak Leaves 100*(23).

L. Ramasubramanian, *Geographic Information Science and Public Participation*, Advances in Geographic Information Science, DOI 10.1007/978-3-540-75401-5,

Bentrup, C. (2002b, June 5). UIC to help with urban planning. *Oak Leaves 100*(27).
Boston Foundation. (2009). *The Boston Indicators Project: About the project*. Retrieved July 1, 2009, from http://www.bostonindicators.org/IndicatorsProject/Content.aspx?id=602
Brail, R., & Klosterman, R. (Eds.). (2001). *Planning support systems: Integrating geographic information systems, models, and visualization tools*. Redlands, CA: ESRI Press.
Brisbane, J. (2005). *PPGIS in action: The Prospect Heights Historic Mapping Project*. Proceedings of the 4th annual public participation GIS conference, Cleveland, OH. Retrieved from URISA, www.urisa.org
Calthorpe, P., & Fulton, W. (2001). *The regional city: Planning for the end of sprawl*. Washington, DC: The Island Press.
Campbell, H., & Masser, I. (1995). *GIS and organizations: How effective are GIS in practice?* London: Taylor & Francis.
Carp, J. (2004). Wit, style, and substance: How planners shape participation. *Journal of Planning Education and Research*, *23*, 242–254.
Chambers, R. (1997). *Whose reality counts: Putting the first last*. Warwickshire, UK: Practical Action Publishing.
Chambers, R. (2002). *Participatory workshops: A sourcebook of 21 sets of ideas and activities*. London: Earthscan.
Chicago Metropolitan Agency for Planning (CMAP), 2007. Public Participation Plan, 2007. Downloadable pdf document, Retrieved July 1, 2009, from http://www.cmap.illinois.gov/
Civic Alliance. (2002). *Listening to the city: Report of proceedings*. New York: Regional Plan Association.
Cooke, B., & Kothari, U. (Eds.). (2001). *Participation: The new tyranny?* London: Zed Books.
Craglia, M., & Masser, I. (2003). Access to geographic information: A European perspective. *URISA Journal*, *15*(1), 51–59.
Craig, W., Harris, T., & Weiner, D. (Eds.). (2002). *Community participation and geographic information systems*. London: Taylor & Francis.
Creighton, J. L. (2005). *The public participation handbook: Making better decisions through citizen involvement*. San Francisco: Jossey-Bass.
Davidoff, P. (1965). Advocacy and pluralism in planning. *Journal of the American Institute of Planners*, *31*(4), 331–337.
Day, D. (1997). Citizen participation in the planning process: An essentially contested concept. *Journal of Planning Literature*, *11*(3), 421–434.
Denzin, N., & Lincoln, Y. (Eds.). (2005). *The sage handbook of qualitative research* (3rd ed.). Thousand Oaks, CA: Sage.
Elwood, S. (2000). *Information for change: The social and political impacts of geographic information technologies*. PhD dissertation, Department of Geography, University of Minnesota. Retrieved from University of Minnesota Libraries.
Elwood, S. (2002). GIS use in community planning: A multidimensional analysis of empowerment. *Environment & Planning A*, *34*, 905–922.
Federal Highway Administration (FHWA). (2007). *The transportation planning process key issues: A briefing book for transportation decisionmakers, officials, and staff*. Published by the Transportation Planning Capacity Building Program, (FHWA-HEP-07-039), Retrieved July 1, 2009, from http://www.planning.dot.gov/documents/BriefingBook/BBook.htm
Foresman, T. (Ed.). (1997). *The history of geographic information systems: Perspectives from the pioneers*. Upper Saddle River, NJ: Prentice Hall.
Forester, J. (1989). *Planning in the face of power*. Berkeley, CA: University of California Press.
Freire, P. (1970). *Pedagogy of the oppressed*. New York: Continuum Publishing.
Friedmann, J. (1987). *Planning in the public domain: From knowledge to action*. Princeton, NJ: Princeton University Press.
Friedmann, J. (1992). *Empowerment: The politics of alternative development*. Oxford: Blackwell.
Gans, H. (1962). *The urban villagers: Group and class in the life of Italian Americans*. New York: Simon & Schuster.

Gaventa, J. (1993). The powerful, the powerless, and the experts: Knowledge struggles in an information age. In P. Park, M. Brydon-Miller, B. Hall, & T. Jackson (Eds.), *Voices of Change: Participatory Research in the United States and Canada* (pp. 21–40). Westport, CT: Bergin & Garvey.

Graham, S., & Marvin, S. (1996). *Telecommunications and the city: Electronic spaces, urban places*. London: Routledge.

Habermas, J. (1987). *A theory of communicative action*. (T. McCarthy, Trans.). Boston: Beacon Press.

Haklay, M., & Tobón, C. (2003). Usability engineering and PPGIS: Towards a user-centred approach. *International Journal of Geographical Information Science (IJGIS)*, *17*(6), 577–592.

Hall, B. (1993). Introduction. In P. Park, M. Brydon-Miller, B. Hall, & T. Jackson (Eds.), *Voices of Change: Participatory Research in the United States and Canada* (pp. xiii–xxii). Westport, CT: Bergin & Garvey.

Hall, P. (1996). *Cities of tomorrow: An intellectual history of urban planning and design in the 20th century*. (Updated edition). London: Blackwell.

Harris, T., & Weiner, D. (1998). Empowerment, marginalization, and "community-integrated" GIS. *Cartography and Geographic Information Systems*, *25*(2), 67–76.

Healy, P. (1996). The communicative turn in planning theory and its implications for spatial strategy formation. *Environment and Planning B: Planning and Design*, *23*, 217–234.

Henkel, H., & Stirrat, R. (2001). Participation as spiritual duty: Empowerment as secular subjugation. In B. Cooke & U. Kothari (Eds.), *Participation: The new tyranny?* (pp. 168–184). London: Zed Books.

Hoch, C. (1994). *What planners do?* Chicago: APA Planners Press.

Hoch, C. (2000). Making Plans. In C. Hoch, L. Dalton, & F. So (Eds.), *The practice of local government planning* (3rd ed., pp. 19–39). Washington, DC: International City/County Management Association.

Hoch, C., Dalton, L., & So, F. (Eds.). (2000). *The practice of local government planning* (3rd ed.). Washington, DC: International City/County Management Association.

Huxhold, W. (1991). *An introduction to urban geographic information systems*. London: Oxford University Press.

Huxhold, W., & Levinsohn, A. (1995). *Managing geographic information systems projects*. Oxford: Oxford University Press.

Innes, J. (1996). Planning through consensus building: A new view of the comprehensive ideal. *Journal of the American Planning Association*, *62*(4), 460–472.

Innes, J., & Booher, D. (2004). Reframing public participation: Strategies for the 21st century. *Planning Theory & Practice*, *5*(4), 419–436.

Irwin, A. (1995). *Citizen science: A study of people, expertise, and sustainable development*. London: Routledge.

Jackson, K. T. (1985). *The crabgrass frontier: The suburbanization of the United States*. New York: Oxford University Press.

Johnson, R. D. (1994). *Where's the power in empowerment? Definitions, differences, and dilemmas of empowerment in the context of work-family boundary management*. PhD dissertation in Organizational Behavior, Harvard University. Retrieved from Harvard Libraries.

Joseph, P. D. (1930). New York city zoning law makes the skyscraper a thing of beauty. *National Civic Review*, *19*(12), 812–814.

Keegan, J. (2006). *Atlantic yards visualization in google earth*. Online resource, Retrieved from: http://invisibleman.com/archives/Atlantic_Yards_Project.kmz

Kennedy, L. W. (1992). *Planning the city upon a hill: Boston since 1630*. Amherst, MA: University of Massachusetts Press.

King, M. H. (1981). *Chain of change: Struggles in Black community development*. Boston: South End Press.

King, M. H., & Ramasubramanian, L. (1997). Research as praxis: The role of an university-based program in facilitating community change. In P. Nyden, D. Burrows, A. Figert, & M. Shibley

(Eds.), *Building community: Social science in action* (pp. 112–119). Thousand Oaks, CA: Pine Forge Press.
Klein, W. (2000). Building consensus. In C. Hoch, L. Dalton, & F. So (Eds.), *The practice of local government planning* (3rd ed., pp. 423–438). Washington, DC: International City/County Management Association.
Korten, D. (Ed.). (1986). *Community management: Asian experience and perspectives*. West Hartford, CT: Kumarian Press.
Kraemer, K., King, J., Dunkle, D., & Lane, J. (1989). *Managing information systems: Change and control in organizational computing*. San Francisco: Jossey-Bass, Inc.
Krieg, R. (1995). Information technology and low-income inner-city communities. *Journal of Urban Technology*, *3*(1), 1–17.
Leitner, H., McMaster, R., Elwood, S., McMaster, S., & Sheppard, E. (2002). Models for making GIS available to community organizations: Dimensions of difference and appropriateness. In W. Craig, T. Harris, & D. Weiner (Eds.), *Community Participation and Geographic Information Systems* (pp. 37–52). London: Taylor & Francis.
Lennertz, B., & Lutzenhiser, A. (2006). *The Charrette handbook: The essential guide for accelerated, collaborative community planning*. Chicago: American Planning Association.
Levine, H., & Harmon, L. (1992). *The death of an American Jewish Community: A tragedy of good intentions*. New York: Simon & Schuster.
Levine, M., & Perkins, D. (1997). *Principles of community psychology: Perspectives and applications* (2nd ed.). New York: Oxford University Press.
Lowry, M., Nyerges, T., & Rutherford, G. S. (2009). Internet portal for participation of large groups in transportation programming decisions. *Transportation Research Record*, *2077*, 156–165.
Lynch, K. (1960). *The image of the city*. Cambridge: MIT Press.
Masser, I., & Onsrud, H. (Eds.). (1993). *Diffusion and use of geographic information technologies*. Dordrecht, Netherlands: Kluwer.
Medoff, P., & Sklar, H. (1994). *Streets of hope: The fall and rise of an urban neighborhood*. Boston: South End Press.
Mitchell, A. (1997). *Zeroing in: Geographic information systems at work in the community*. Redlands, CA: ESRI Press.
Mitchell, W. (1995). *City of bits. Space, place, and the infobahn*. Cambridge, MA: MIT Press.
Mitchell, W. (1999). The digital revolution. In D. Schön, B. Sanyal, & W. Mitchell (Eds.), *High technology and low-income communities: Prospects for the positive use of advanced information technology* (pp. 105–130). Cambridge, MA: MIT Press.
Mollenkopf, J. H. (1983). *The contested city*. Princeton, NJ: Princeton University Press.
Mossberger, K., Tolbert, C., & McNeal, R. (2008). *Digital citizenship: The internet, society and participation*. Cambridge, MA: MIT Press.
Mossberger, K., Tolbert, C., & Stansbury, M. (2003). *Virtual inequality: Beyond the digital divide*. Washington, DC: Georgetown University Press.
Myers, J., Martin, M., & Ghose, R. (1995). GIS and neighborhood planning: A model for revitalizing communities. *URISA Journal*, *7*(2), 63–67.
Naisbitt, J. (1994). *The global paradox: The bigger the world economy, the more powerful its smallest players*. New York: Avon Books.
Negroponte, N. (1995). *Being digital*. New York: Alfred A. Knopf.
Norris, P. (2001). *Digital divide: Civic engagement, information poverty, and the internet worldwide*. Cambridge: Cambridge University Press.
Oak Park. (1990). Comprehensive Plan. Retrieved from http://www.oak-park.us/Commissions/Plan_Commission.html
Obermeyer, N. (1998). The evolution of public participation GIS. *Cartography and GIS*, *25*(2), 65–66.
Obermeyer, N., & Pinto, J. (2008). *Managing geographic information systems*. 2nd edition. New York: Guilford Press.
Onsrud, H., & Pinto, J. (1993). Correlates of GIS adoption success and the decision process of GIS acquisition. *URISA Journal*, *4*, 32–44.

Ozawa, C. (1991). *Recasting science: Consensual procedures in public policy making*. Boulder, CO: Westview Press.

Park, P. (1993). What is participatory research? A theoretical and methodological perspective. In P. Park, M. Brydon-Miller, B. Hall, & T. Jackson (Eds.), *Voices of Change: Participatory Research in the United States and Canada* (pp. 1–20). Westport, CT: Bergin & Garvey.

Peterman, W. (1996). The meanings of resident empowerment: Why just about everybody thinks its a good idea and what it has to do with resident management. *Housing Policy Debate*, *7*(3), 473–490.

Pickles, J. (Ed.). (1995). *Ground truth: The social implications of Geographic Information Systems*. New York: The Guilford Press.

Poiker, T. K. (1976). A theory of the cartographic line. *International Cartographic Yearbook*, *16*, 134–143.

Ramasubramanian, L. (1995). Building communities: GIS and participatory decision-making. *Journal of Urban Technology*, *3*(1), 67–79.

Ramasubramanian, L. (1998). *Knowledge production and use in community-based organizations: Examining the impacts and influence of information technologies*. PhD dissertation, School of Architecture and Urban Planning, University of Wisconsin-Milwaukee (UWM). Retrieved from the UWM Libraries.

Ramasubramanian, L. (1999). Nurturing community empowerment: Participatory decision-making and community-based problem solving using GIS. In H. Onsrud and M. Craglia (Eds.), *Geographic Information Research: Transatlantic Perspectives* (pp. 87–102). London, England: Taylor & Francis.

Ramasubramanian, L. (2004). Knowledge production and use in community-based organizations: The impacts and influence of information technologies. *CityScape*, *7*(1), 165–191.

Ramasubramanian, L. (2007). Access to geographic information. In K. Kemp (Ed.), *Encyclopedia of Geographic Information Science* (pp. 1–2). London: Sage.

Rambaldi, G., Kwaku-Kyem, P., Mbile, P., McCall, M., & Weiner, D. (2006). Participatory spatial information management and communication in developing countries. *The Electronic Journal of Information Systems in Developing Countries (EJISDC)*, *25*(1), 1–19.

Rheingold, H. (1993). *The virtual community: Homesteading on the electronic frontier*. Cambridge, MA: MIT Press.

Rietbergen-McCracken, J., & Narayan-Parker, D. (1998). *Participation and social assessment: Tools and techniques*. Washington, DC: The World Bank.

Rivera, F., & Erlich, J. (1992). *Community organizing in a diverse society*. Boston: Allyn and Bacon.

Rivera, R. (2007, January 18). Plan will allow 911 and 311 lines to accept digital images. *New York Times*. Retrieved from www.nytimes.com

Rocha, E. (1997). A ladder of empowerment. *Journal of Planning Education and Research*, *17*, 31–44.

Rogers, E. (1983). *Diffusion of innovations*. New York: The Free Press.

Rogers, E. (1995). *The diffusion of innovations* (4th ed.). New York: The Free Press.

Sawicki, D., & Peterman, D. (2002). Surveying the extent of PPGIS practice in the United States. In W. Craig, T. Harris, & D. Weiner (Eds.), *Community Participation and Geographic Information Systems* (pp. 17–36). New York: Taylor & Francis.

Schlossberg, M., & Shuford, E. (2005). Delineating "public" and "participation" in PPGIS. *URISA Journal*, *16*(2), 15–26.

Schön, D. (1983). *The reflective practitioner*. New York: Basic Books.

Schön, D., & Rein, M. (1994). *Frame reflection: Toward the resolution of intractablepolicy controversies*. New York: Basic Books.

Schuler, D. (1997). *New community networks: Wired for change*. New York: Addison-Wesley Publishing Company

Seiber, R. (2006). Public participation GIS: A literature review and framework. *Annals of the Association of American Geographers*, *96*(3), 491–507.

Shaw, D., & Sykes, O. (2005). Addressing connectivity in spatial planning: The case of the English regions. *Planning Theory & Practice, 6(1)*, 11–33.

Shenk, D. (1997). *Data smog: Surviving the information glut*. New York: HarperCollins Publishers.

Shiffer, M. (1995). Interactive multimedia planning support: Moving from stand-alone systems to the World Wide Web. *Environment and Planning B: Planning and Design*, *22*, 649–664.

Snow, C., Pettit, K., & Turner, M. (2004). *Neighborhood early warning systems: Four cities' experience and implications for the District of Columbia*. Final report prepared by the Urban Institute's Metropolitan Housing and Communities Policy Center for the Fannie Mae Foundation, March 2004. Retrieved from the Fannie Mae Foundation.

South End Housing and Planning Coalition. (1991). Boston, *A biomedical frontier: Hype or hope*. Boston: SEPHC Publication.

Squires, G. (Ed.). (1992). *From redlining to reinvestment: Community responses to urban disinvestment*. Philadelphia, PA: Temple University Press.

Squires, G., Bennett, L., McCourt, K., & Nyden, P. (1989). *Chicago: Race, class, and the response to urban decline*. Philadelphia: Temple University Press.

Stephenson, R. (1998). In what way, and to what effect is technical information used in policy making? Findings from a study of two development plans. *Planning, Practice & Research*, *13*(3), 237–245.

Stoll, C. (1995). *Silicon snake oil: Second thoughts on the information highway*. New York: Doubleday.

Susskind, L., Elliott, M., & Associates. (1983). *Paternalism, conflict, and co-production: Learning from citizen action and citizen participation in Western Europe*. New York: Plenum Press.

Talen, E. (2000). Bottom-up GIS: A new tool for individual and group expression in participatory planning. *Journal of the American Planning Association*, *66*(3), 279–294.

Throgmorton, J. A. (2003). Planning as storytelling in a global-scale web of relationships. *Planning Theory*, *2*(2), 125–151.

Tomlinson, R. F. (1987). Current and potential uses of geographical information systems, The North American experience. *International Journal of Geographical Information Systems*, *1*, 203–218.

US Supreme Court. (1926). Village of Euclid, Ohio v. Ambler Realty Co., 272 U.S. 365. Online resource, Retrieved from http://caselaw.lp.findlaw.com/scripts/getcase.pl?court=US&vol=272&invol=365

US Supreme Court. (2005). Kelo et al. v City of New London et al., No. 04-108. Online resource, Retrieved from http://caselaw.lp.findlaw.com/scripts/getcase.pl?court=US&vol=000&invol=04-108

Village of Oak Park. (2004). Minutes of the regular meeting of the President and board of trustees of the Village of Oak Park, held on Monday, March 15, 2004 at 7:30 p.m. in the council chambers of the village hall, Retrieved July 1, 2008, from http://www.oakpark.us/About_Our_Village/board_minutes_2004.htm

Village of Oak Park. (2005). Diversity Statement adopted by the Village President and Board of Trustees on May 2, 2005. Retrieved July 1, 2008, as a downloadable pdf file from http://www.oak-park.us/Living_In_Oak_Park/Living_In_Oak_Park.html

Village of Oak Park. (2006). *Village Board of Trustees Policy, February 2006: Public participatory planning guidelines for village owned properties*. Retrieved July 1, 2008, from http://www.oak-park.us/Community_Services/Planning.html

Village of Oak Park. (2007). *Village background*. Retrieved July 1, 2008, from http://www.oak-park.us/Village_Background/Village_Background.html

Voogd, H., & Woltjer, J. (1999). The communicative ideology in spatial planning: Some critical reflections based on the Dutch experience. *Environment and Planning B: Planning and Design*, *26*(6), 835–854.

Wegener, M., & Masser, I. (1996). Brave new GIS worlds. In I. Masser, H. Campbell, & M. Craglia (Eds.), *GIS Diffusion: The adoption and use of Geographical Information Systems in local government in Europe*. London: Taylor & Francis.

Weir, M. (2000). Coalition building for regionalism. In B. Katz (Ed.), *Reflections on Regionalism*. Washington, DC: Brookings.

Werner, A. (2004). *A guide to implementation research*. Washington, DC: The Urban Institute Press.

Wheeler, S. (2002). The new regionalism: Key characteristics of an emerging movement. *Journal of the American Planning Association*, *68*(3), 267–278.

Wiedemann, P. M., & Femers, S. (1993). Public participation in waste management decision making: Analysis and management of conflicts. *Journal of Hazardous Materials*, *33*(3), 355–368.

Witten, K., Parkes, M., & Ramasubramanian, L. (2000). Participatory environmental health research in Aotearoa/New Zealand: Constraints and opportunities. *Health Education and Behavior*, *27*(3), 371–384.

Wright, D., Goodchild, M. F., & Proctor, J. D. (1997). GIS: Tool or science? Demystifying the persistent ambiguity of GIS as "Tool" versus "Science". *Annals of the Association of American Geographers*, *87*(2), 346–362.

Whyte, W. (Ed.). (1991). *Participatory action research*. Newbury Park, CA: Sage.

Yankelovich, D. (1991). *Coming to public judgment*. Syracuse, NY: Syracuse University Press.

Yaro, R., & Hiss, T. (1996). *A region at risk: The third regional plan for the New York-New Jersey-Connecticut metropolitan area*. Washington, DC: Island Press.

All web links were current when the book went into publication. Please report problems with missing or broken links to the author by writing to: Prof. Laxmi Ramasubramanian, Department of Urban Affairs and Planning, Hunter College, 695 Park Avenue, New York, NY 10065.

About the Author

Laxmi Ramasubramanian, PhD., is an associate professor in the Department of Urban Affairs and Planning at Hunter College and Visiting Professor (Overseas) at Anna University, in Chennai, India. Ramasubramanian is trained as an architect and a city planner. After completing her undergraduate and graduate education in architecture in India, Dr. Ramasubramanian moved to the United States where she received her Master in City Planning degree from the Massachusetts Institute of Technology. Her PhD research at the University of Wisconsin-Milwaukee examined how digital technologies influence the work of community-based organizations. Ramasubramanian has previously held research and teaching appointments at the University of Illinois-Chicago, the University of Wisconsin-Milwaukee as well as postdoctoral fellowships at the University of New England, NSW, Australia and the University of Auckland, New Zealand.

Dr. Ramasubramanian's research examines how the use of digital technologies such as GIS alters social and political processes, particularly the power of individuals and institutions to create and sustain social change. Her research agenda cuts across conventional disciplinary boundaries. She works collaboratively with engineers, computer scientists, and public health professionals. Dr. Ramasubramanian is an expert in the design, implementation and evaluation of participatory planning projects that use affordable and accessible digital technologies. Professor Ramasubramanian's research has uncovered that technology use by relatively powerless actors (community-based organizations, for example) alters the nature of knowledge creation and thereby the discursive strategies they can use to solve strategic planning problems.

Dr. Ramasubramanian has received research funding from a range of public and nonprofit sources including the US Department of Housing and Urban Development, US Federal Transit Administration, NOAA, the Ford Foundation, the NSF-funded National Center for Geographic Information and Analysis, and the New York Metropolitan Transportation Council. Professor Ramasubramanian has been active in research and project management, serving as Principal Investigator or Co-Principal Investigator of 18 research projects totalling over

2 million dollars over the past 8 years. She currently serves on the board of directors of the University Consortium for Geographic Information Science (UCGIS), and is active in the Urban and Regional Information Systems Association (URISA).

Dr. Laxmi Ramasubramanian
Department of Urban Affairs and Planning
Hunter College
695 Park Avenue, Room 1645 HW
laxmi [at] hunter.cuny.edu
New York, NY 10065

Index